KU-325-317

WEAPONS

OF MASS PERSUASION

Lynda Lee Kaid and Bruce Gronbeck
General Editors

Vol. 15

PETER LANG
New York • Washington, D.C./Baltimore • Bern
Frankfurt am Main • Berlin • Brussels • Vienna • Oxford

WEAPONS

OF MASS PERSUASION

Strategic Communication
to Combat Violent Extremism

EDITED BY
Steven R. Corman, Angela Trethewey,
& H. L. Goodall, Jr.

PETER LANG
New York • Washington, D.C./Baltimore • Bern
Frankfurt am Main • Berlin • Brussels • Vienna • Oxford

Library of Congress Cataloging-in-Publication Data

Weapons of mass persuasion: strategic communication to combat violent extremism /
edited by Steven R. Corman, Angela Trethewey, H.L. Goodall, Jr.
p. cm.—(Frontiers in political communication; v. 15)
Includes bibliographical references and index.
1. Communication in politics—United States. 2. Terrorism and mass media.
3. National security. 4. Communication, International.
5. Mass media and public opinion. I. Corman, Steven R.
II. Trethewey, Angela. III. Goodall, H. Lloyd.
JA85.2.U6W43 320.97301'4—dc22 2008002470
ISBN 978-1-4331-0198-4 (hardcover)
ISBN 978-1-4331-0197-7 (paperback)
ISSN 1525-9730

Bibliographic information published by **Die Deutsche Bibliothek**.
Die Deutsche Bibliothek lists this publication in the "Deutsche
Nationalbibliografie"; detailed bibliographic data is available
on the Internet at http://dnb.ddb.de/.

Cover design by Joni Holst
Front cover image is the logo for the
Consortium for Strategic Communication (www.comops.org)

The paper in this book meets the guidelines for permanence and durability
of the Committee on Production Guidelines for Book Longevity
of the Council of Library Resources.

29 Broadway, 18th floor, New York, NY 10006
www.peterlang.com

Printed in the United States of America

Contents

Part II: Case Studies on Communication, Terrorism, & National Security

Part III: Crafting a New Communication Policy

Preface

STEVEN R. CORMAN, ANGELA TRETHEWEY,
AND H. L. GOODALL, JR.

You may wonder, as we three editors of this volume often do, how otherwise ordinary American academics from the Communication field got involved in the so-labeled "Global War on Terror?" What caused us to move from the safety and relative security of our university-sculpted tenured lives into the ongoing conversations about combating ideological support for terrorism, the role of communication in public diplomacy, and other questions about this new "rugged terrain" of fear, danger, lies, death, and loathing?

Here is the short version of what happened. In 2005, under the direction of Steven R. Corman, we formed the Consortium for Strategic Communication[1] and began edging our way into the ongoing public conversations by publishing white papers in response to ongoing blunders and missteps that seemed to us to require, rather urgently, an informed response from the scholarly world of Communication studies. It seemed entirely fitting for those of us who have devoted lifetimes of study to strategic communication, intercultural communication, organization and network analysis, and rhetoric to provide input to a war of ideas that we were quickly losing.

At first we were convinced, as are most Americans, that *someone* must have access to the same scholarly information that we have. After all, communication research isn't classified and even the most basic of textbooks in the field cover topics that seem to have eluded decision makers inside the beltway. For example, we could not believe a President of the United States would want to use an internationally broadcast address to patently offend the world's largest religion by declaring a "crusade" in the Middle East. Yet he did so. Similarly, we questioned the use of the rhetorically loaded term "war" to describe the collective international efforts at combating terrorism on grounds that would be familiar to anyone who studies the power of narratives to shape the public mind and opinion. And there seemed to be a disconnect theoretically as well as practically in the government and military understanding of the role of new media in radical extremist recruitment and disseminating information to terrorist cells.

As we considered the challenges and our options, the issues continued to accumulate. Nothing new from inside the beltway seemed to emerge. Hmmmmm.

Similarly, the engagement of fellow academics seemed limited to scathing critiques of administration policies and the foibles of various spokespersons. As academics we all firmly believe that critique is critical to understanding, and that theoretical advancements are built upon a solid foundation of criticism of existing ways of thinking. But we were also aware of the vast gap between academic critiques of the global war on terror and the practical considerations concerning communication practices of those charged with waging it. We hoped that our white papers would find an audience among the latter group and that from that prose introduction we would engage in some productive conversations aimed at changing the present course of applied communication in the global war of ideas.

This goal also fit our academic mission. As scholars at Arizona State University, a "New American University" under the vision and leadership of President Michael C. Crow, we are charged with using our academic home as a think tank for use-inspired research that engages and informs pressing social problems.

One tenet of that vision is the freedom to create interdisciplinary and interorganizational collaborations. So it was under the auspices of the Hugh Downs School of Human Communication that we created a public lecture series on "Communication, Terrorism, and National Defense," formed a strong working relationship with the Center for Combating Terrorism at West Point and the

Center for the Study of Religion and Conflict at ASU, and began publishing white papers about the relationship of communication to the war of ideas and how it might be better be waged.[1]

Our work attracted the attention of the Special Operations Command, the U.S. State Department, and the U. S. Department of Defense. Members of our Consortium have been invited to national and international conferences to air our views, and to share our understandings with those actually waging this war both in the field and on the Internet. Peter Singer of the Brookings Institute, writing an op-ed for the *Washington Post*, reviewed our "Pragmatic Complexity" white paper as an "important new study," highlighting its direct relevance to countering the current failed communication efforts and offering a new model for leadership capable of reversing current practices. (http://www.washingtonpost.com/wpdyn/content/article/2007/05/15/AR2007051500995.html)

Our new scholarly efforts represent a "system perturbance" aimed at disrupting the way broader audiences conceive of, and use, communication in this, the most pressing communication issue of the 21st century. It is important and practical work. Our work is designed to reach broader public audiences, decision makers, and policy groups without giving up the cultural and critical foundations central to our scholarship. We have been, therefore, surprised and delighted to find that what advice and criticism we can offer based on our theoretical knowledge and pragmatic orientation to the conduct of communication in public diplomacy and counter-terrorism has been well received by these diverse audiences.

We have also been summarily disappointed to realize that what we fear to be true is, in fact, true: Our current Bush administration and its political appointees are failing to win the war of ideas (as well as the war in Iraq and Afghanistan) because they seem to believe that they have always been right. They have not, until quite recently when so much ground and so many lives have been sacrificed, been open to new ideas about how they communicate with the world nor have they been receptive, much less responsive, to criticism. That posture—their view of how to "do" communication—must change if we are to regain our stature in the world community.

All change begins in conversation and all significant diplomatic accomplishments are *rooted in communication strategies*, some of which are obvious (open and honest dialogue) and others that are, in fact, counter-intuitive. From a communication perspective, the failure of this administration's leaders and spokespersons to embrace contemporary communication theories and practices provides both a source of critique as well as an opening for new conversations that educate those who are in positions to change how strategic communication is accomplished.

This war of ideas is not, to borrow Samuel Huntington's ill-chosen phrase a "clash of civilizations" but instead *a clash of systems and cultures of communication* that are themselves firmly rooted in diverse religious, economic, and social conditions. What is needed is not a rhetoric of further divisiveness that deepens the conflict or a stubborn insistence on old rules for the conduct of public diplomacy, but instead a leadership capable of demonstrating our most cherished democratic values by engaging honestly with others, by speaking to and (at least as important) listening to those who disagree with our policies and providing them with opportunities to talk through our differences.

There will, of course, be differences that we cannot talk through and groups that we cannot talk to. That is, at this point in the evolution of this war, inevitable. Similarly, there are others—terrorists and other violent extremists who have joined the global social movement against us—who neither want to engage us in meaningful dialogue nor will settle for anything less than our total destruction. For these others, only continued use of military solutions will bring a resolution of our differences. That is unfortunate, but it is also true.

However, dealing with violent extremists who are willing to strap bombs on their bodies or sponsor the destruction of innocent lives using whatever weapons of mass destruction they may acquire is *not* where the war of ideas is being waged nor the audience for the messages we need to deploy. Our country needs to appeal to those who once believed that America was a beautiful idea and a force for good in the world, and who may yet be persuaded of that fact again. We must also seek new channels of communication that depart from traditional approaches and venues for public speeches and forums. We must learn to leverage all that we know about new media, viral marketing, gaming, virtual reality, and cultural diplomacy to once again offer hope to a troubled world and practical solutions to everyday problems of food, shelter, safety, and security in the world's troubled neighborhoods. Moreover, we need to find pathways to the minds of those who may still be persuaded that America is not the Great Satan, that freedom is not corrupt, and that democracy is capable of working within existing systems of religion and culture.

The "war of ideas" is a communication struggle. It cannot be won militarily on the battlefield but must be won rhetorically and narratively in the hearts and in the minds of those on all sides of this ideological front who can—who must—come to believe that finding a better way of respectfully exchanging views is preferable to finding better ways of destroying each other.

That said, there is no doubt that the radical extremists who want to destroy us must be met in kind. But they are not our audience, nor should they be conceived of as the audience for a policy of strategic communication engagement. They are, at best, criminals and at worst madmen who must be brought to justice by any means possible and as quickly as possible.

Our audience for strategic communication consists of those peace-loving but conflicted souls who may yet be infected by the spreading virus of a violent extremist's ideological message as well as those spokespersons and policy makers on our side who make strategic plans to find and to influence them. These are admittedly extremely disparate audiences, yet they are unified by world events that have not yet been leveraged as communication opportunities capable of creating common interests capable of transforming our differences.

The chapters that comprise this volume attempt to take a giant step in that specific direction. They demonstrate the practical utility of use-inspired research that applies contemporary theories of communication and media to real-world problems of countering ideologically sanctioned terrorism and, we hope, promote a better vision for the conduct of public diplomacy.

Chapter 1 provides a critical reading of the faulty assumptions and failed communication strategies resulting from an outmoded theoretical orientation toward strategic communication by the Bush administration, the Pentagon, and various government agencies charged with winning the hearts and minds of global audiences. The remaining chapters address these communication failures by applying contemporary theoretical understandings about communication to specific cases in the global struggle against violent extremism. Chapter 2 applies the principle of *strategic ambiguity* to leadership and public diplomacy challenges in the Middle East. In Chapter 3 we reveal a principle of *fundamentalist rhetoric* that enabled the dark side of leadership to emerge and flourish since 9/11 in ways that closely resemble similar patterns during the Cold War.

The fourth chapter unveils the communication principles that drive violent extremists' media strategies and that offer insights into how allied powers must expose and counter violent extremists on the Internet. Chapter 5 examines President Mahmoud Ahmadinejad's letter to President George W. Bush within the context of jihadi media strategies and our ineffective responses to them. Similarly, Chapter 6 demonstrates the importance and strategic value of understanding the principle of *visual messages on diverse global audiences* as a key element in winning the war of ideas. Chapter 7 explores the principle of *language choice, naming* and the deployment of terminology as key to an informed strategic communication plan.

Chapter 8 offers a bold new theoretical orientation for strategic communication in the war of ideas. Grounded in the principles from the preceding chapters, the theory of *pragmatic complexity*, reveals new opportunities to rethink strategic communication policies and practices. In Chapter 9 we conclude the volume by rewriting the current U. S. strategic communication plan to incorporate the principles contained in this volume. Our hope is that policy makers, scholars, and practitioners will benefit from these insights. More importantly, we hope that these insights provide real world applications capable of reversing the negative image of the United States in world affairs and once again demonstrating the value of enlightened leadership and informed communication practice in the global struggle against violent extremism.

Steven R. Corman, Angela Trethewey, and H. L. Goodall, Jr.
Tempe, Arizona
December 2007

Note

1. See our website (http://www.asu.edu/clas/communication/about/csc/)

PART I

Strategic Communication in a Post-9/11 World

CHAPTER ONE

"Strategery"[1]

Missed Opportunities and the Consequences of Obsolete Strategic Communication Theory

H.L. GOODALL, JR., ANGELA TRETHEWEY, AND STEVEN R. CORMAN

Background

There are two major trends that underscore the need for improved strategic communication efforts in the war of ideas.[2] First, we have witnessed a steady increase in international terrorism, suicide bombings, and the spread of violent extremist ideologies despite a sustained military and diplomatic assault by the United States, Great Britain, and a small number of allies, all under the label of fighting a "global war on terror" (*Militant Ideology Atlas*, November 2006). Soumaya Ghannoushi, director of research at IslamExpo, traces the growth and expansion of terrorist cells:

> *When Osama bin Laden and his lieutenant Ayman al-Zawahiri issued their "Jihad against Jews and Crusaders" statement on February, 28 1998, responses to their declaration varied from apathy to amusement. They were an obscure group lost in the faraway emirate of the Taliban, a pathetic remnant of the fight against the USSR during the cold war. Their role looked historically defunct and their discourse archaic.*

> *Things could not be more different now. Al-Qaida has become an intensely complex global network, with a decentralised, flexible structure that enables it to spread in all directions, across the Arab world, Africa, Asia and Europe. Whether pursuing active cells or searching for sleeping ones, the security world is haunted by al-Qaida's ghost. Like bubbles, these cells are autonomous, bound together neither by hierarchy nor by a chain of command. It only takes a few individuals who subscribe to its ideology and terrorist methods for al-Qaida to extend its reach to a new part of the globe.*
>
> *With the Middle East moving from one crisis to another, this small organisation saw itself miraculously transferred from periphery to centre. In its founding statement, al-Qaida defined its mission as a jihad aimed at cleansing the Arabian peninsula of the American "locusts, eating its riches and wiping out its plantations," and liberating Palestinian land from Zionist occupation. With the invasion of Iraq in 2003, al-Qaida was offered a firm foothold in the Middle East and the unique chance to implement its "resistance against Jews and crusaders" project. . . .*
>
> *Rapidly expanding from one location to another, al-Qaida currently boasts branches throughout the Arab region. These developments are worrying not only from the point of view of ruling governments and their western allies, but from that of mainstream Islamic movements too.*[3]

The Middle East is not the only host to terrorist cells. "Homegrown" terrorist cells that are either inspired by bin Laden, if not directed by him, have emerged across Europe, Asia and Africa.

Second, there has been a steady decline of America's image abroad that has also brought with it a steady decline in worldwide support for our efforts, policies, and interventions in the Middle East and elsewhere (*Pew Global Attitudes Report*, June 13, 2006). Taken together, these disturbing trends point to *the most dramatic failure of the United States to plan, coordinate and execute a successful strategic communication campaign in the history of our nation.*[4]

A communication strategy alone cannot overcome faulty decisions and problematic policies. However, one clearly consistent feature of U. S. communication practice in the Global War on Terror (GWOT) is a *heavy reliance on an antiquated, linear, and simplistic model of communication to guide public diplomacy and counter-terrorism efforts*. This model, drawn from a public influence approach to selling messages as if they were products to receptive and passive consumers who share the same values, fails to adequately account for the globalized, complex, media-saturated and ideologically-laden environment in which the meanings of messages and messengers are interpreted (For details of the policy, please see *U. S. National Strategy for Public Diplomacy and Strategic Communication*, December 2006).

Just as we failed to adequately "connect the dots" leading up to 9/11[5], since

9/11 we have failed to read and deal with responses to our messages from those we most hope to influence. That failure is, again, tied to an outmoded and inappropriate model of communication that is poorly adapted to the task of persuading audiences already *hostile to the messenger* and *disenchanted with the message.* For example, President Bush and Secretary of State Rice simply declared freedom and democracy to be universally embraced values without a full understanding of how those terms are made meaningful by targeted audiences. Not only were they not ideal spokespersons (because they lack credibility with the intended audience), but the message itself was often fraught with *strategery:* soundbite statements that might play well in the U. S., but are interpreted in radically different ways among diverse global audiences. As a result, the messengers were discredited and the message was either discounted, distorted, and/or used as propaganda for the audiences we attempted to persuade. The unreflective use of potentially equivocal or even inflammatory terms—such as freedom and democracy—coupled with a simplistic public influence model of communication that foregrounds the sender's message instead of the audience's interpretations of those messages characterizes our largely unsuccessful strategic communication efforts since 9/11.[6]

Failures to Communicate: Four Missed Opportunities

The meaning and importance of communication are largely in the minds and hearts of those *on the receiving end of it.* This basic assumption of communication effectiveness has been largely absent in the public strategic communication policies and actions of the Bush administration since 9/11. Both public diplomacy and counter-terrorism efforts have been based on public relations and advertising practices that do not heed recent advances in communication theory. One great lesson has been learned: you can't sell complex concepts such as freedom, democracy, equality, or even the *idea* of America in the same way as Western marketers sell prepackaged rice.

Here's why. For years, communication scholars generally, and intercultural communication scholars particularly, have moved away from a "linear" model of communication as a one-way, message-driven, branding campaign. This antiquated model, derived initially from the Shannon and Weaver conception of how to reduce noise in analog telephones in 1948—and afterward refined by public relations and marketing firms to sell products using simplistic slogans, repetitive messages, and market saturation strategies—relies on the premise that communication is largely information transfer controlled by the sender whose messages act as persuasive sources of motivation on passive audiences.[7] This model is fundamentalist in the sense that it assumes that strategic communicators are essentially engaged in a quest to locate a simple sacred message made up of a few holy words or "god terms"[8] capable of being instantly transferred from the mind of a dominant, all-knowing sender into the minds of multiple listeners. Accordingly, that assumption is that listeners will automatically understand and interpret the holy words in the same way intended by the sender. Listeners are conceived as passive receptors who cannot doubt the veracity of the message: There is no room for alternative interpretations, and no allowance made for differences in points of view, the nuances of culture, the doctrines of different religions, or the unique minds of women and men. All the better if they do not entertain alternative points of view out of fear.[9] This "fundamentalist" model obviously works well for al Qaeda and other violent extremist groups, but as a model of communication intended for those who are not already "true believers" outside of American culture, it leaves a lot to be desired.[10]

By contrast, over the past several decades communication scholars have advanced theory in a very different direction.[11] We have moved toward a more complex appreciation and understanding of the communication process, as one that is always audience based, culturally dependent, and meaning-centered. Messages are never disconnected from the *ongoing narrative stream* that informs, surrounds and constitutes them.[12] In this way, communication is understood as a mutual construction of meaning, or, if you prefer, as a tenuous joint venture informed as much by culture, ambiguity, silences, metaphor, and activity as by the clarity of a campaign of holy words, the depth of market saturation, or the righteousness of intentions.[13]

What is needed in the Global War on Terror (beyond a renaming of it; see Chapter 7 of this volume) is an understanding of communication as the ability of actors who *do not* share meanings, religions, political views, or understandings of history, to act in the world *collaboratively despite their differences*. Viewed from this alternative perspective, communication is always going to be an *ongoing narrative process of sensemaking*[14] that is heavily dependent for meaning on mediated perceptions of current events, feelings about prior interactions, existing interpretive schemes, the influence of social networks, identification with cultural/political/religious identities, and the resulting taken-for-granted norms and expectations for conduct, for reasoning, and for moral judgment.

Communication is the *material manifestation of consciousness*—the outward performance of a cultural and spiritual nexus.[15]

One further tenet of the narrative communication process is that there is *nothing neutral* about the information content of the messages that comprise it. This tenet is important for a number of reasons. First, it means that all messages are partial, partisan, and problematic because the information contained within them is always rich with cultural and political interpretation and perspective.[16] Second, it means that the achievement of meaning and/or the coordination of action among disparate audiences is never a simple matter of making linguistic or grammatical correspondences between symbols or signs to issues of "fact." There are *no culture facts apart from the narratives that structure and give them relevance* to the meaning of the current situation. And finally, because there is only ever an *interpretation* of the facts and interpretations of facts are deeply rooted in core cultural narratives, not even a clear and seemingly compelling disparity between known facts and demonstrable realities likely has sufficient rhetorical power to overturn them.

Not only arc there no meaningful facts independent of local interpretations, but neither can the meaning of a message be controlled to any great degree by the sender of it. In fact, senders have far *less* power to determine meaning than do those whose local work is making sense of it within their own interpretive frameworks and narrative streams. That is because those who *receive* and make sense of messages have the power—indeed, the imperative—to act on what they *perceive* the "true" meaning (as well as its "intended" meaning) to be.

The real work of communicating with disparate others, then, is not in finding "just the right message" or in saturating a targeted audience with it. Instead, the real work of human communication is in trying to find *newer, better ways of organizing resources at all levels of culture, language, and strategy* capable of influencing how messages are likely to be received and interpreted in an ever-changing and evolving, complex and meaning-centered narrative world. The resources are partly based in strategic choices of language to work within and, possibly, to alter existing narrative frameworks; partly in making productive use of cultural and religious/spiritual understandings; and partly in *not doing what is expected*, particularly in times of seemingly intractable conflict. [17]

The chapters in this book are informed by that singular message, and the larger narrative we have crafted out of communication and cultural theories. It is a message that allows us, to function as applied communication critics who examine recent past events critical to the diplomatic, political, and military missions involved in the struggle against violent extremism. It also allows us to use the missteps that characterize those events to focus on real world correctives—what should have been said or done—in order to inform future communication actions, policies, and plans. Scenarios, like historical bad habits, do repeat themselves unless new ways of thinking provide the impetus to change them.

In our view, our government has failed, however unwittingly, to be sensitive to communication practices that mark their public pronouncements, information campaigns, and policies. In the following section we briefly explain four major themes that demonstrate missed opportunities for transformative communication, and foreshadow the major arguments that are contained throughout the case studies presented later in this volume. Those themes center on (1) a simplistic "soundbite" vocabulary to narrate the struggle against violent extremism; (2) a sender-oriented approach to public diplomacy; (3) lost opportunities for repairing U. S. credibility and image across the globe; and, (4) the failure to fully engage local narratives and networks as strategies for countering ideological support for terrorism.

Soundbite Strategery: The Need for a New Vocabulary

Much has already been written on the problematic vocabulary of the misnamed War on Terror.[18] Why "misnamed?" Because, from a communicative perspective, wars are fought against soldiers of opposing nations, not against nouns. Previous wars on nouns—the war on drugs, the war on illiteracy, the war on poverty—failed for the same reason a war on terror will fail: *you cannot defeat that which cannot be properly named*. Calling something a "war" taps into well-established cultural narratives that begin with an event and end with an event; they do not go on forever. Therefore, trying to rally narrative resources around an "enduring war," or a "long, hard slog" is inherently confusing to audiences at home, and our allies, who have been reared culturally on Western war stories.[19]

Rather than repeating these well-known arguments, we concentrate instead on the relationship of strategic language to larger issues of framing. For illustration, we use the linguistic choices made by President Bush that were designed to win support for his vision and policies for the "War on Terror."[20] In the days and weeks after 9/11, President Bush's message was one that emphasized that America was now "at war," that the nations of the earth were either "with us or with the terrorists"[21] that our goal was "to bring terrorists to justice or bring justice to the terrorists" that there was no difference between terrorists and those who harbor them, and that "freedom" would triumph over "fear."

These and subsequent statements established President Bush as the principal spokesperson for the United States and its allies in what would become known as the Global War on Terror. No doubt, the advice that he was given emphasized the need to appear strong, decisive, and inspirational for audiences in our culture and throughout the Western democracies who wanted him to make sense of an uncertain and anxiety-producing context and to offer up assurances that good would triumph over evil.[22]

Furthermore, the man given the task of defining the war was, himself, a cultural product of American war narratives, especially FDR's response to Pearl Harbor. This no doubt figured into Bush's decision to make his first televised address to the American people from behind FDR's desk. Together this President and his American audience craved a simple unifying message capable of reducing our anxieties while rhetorically countering the tragic events of 9/11 with weapons of mass persuasion.[23]

What President Bush and his advisors did not appreciate was the depth of the existing anti-Western culture in the Middle East, the extreme anti-American interpretive frameworks, and rhetoric for global *jihad* that dominates Islamic resistance organizations such as al Qaeda, Hamas, Hezbollah, and others.[24] But more importantly, for audiences of the middle range—composed of soft opposition, undecided, and soft supporters of Western ideas and interests—the President's initial message and those following it were alarming in their totalizing characterization of Islam and its use of "crusade" rhetoric.[25]

But perhaps the largest strategic error evident in these messages was how cleanly they played into the hands of Osama bin Laden and his supporters.[26] By striking America hard enough to cause troops to be deployed in Afghanistan and later Iraq, the American President fulfilled bin Laden's prophesies and validated the warnings he and other extremist leaders had been preaching for the past decade. Senior intelligence officials, State Department diplomats, and former President Bill Clinton had warned the White House of these predictions, of course, but neither the President, nor the Vice President, nor the Secretary of Defense, nor the Director of the CIA, nor the National Security Advisor heeded their warnings.[27]

Viewed from the perspective of the average Islamic citizen in the Middle East, Indonesia, Asia, Europe, and North and South America, the President's use of "crusade" rhetoric validated fears stoked by the extremist leaders: that the United States, a nation led by a avowed, born-again, self-described fundamentalist Christian and addicted to oil, would use *any* excuse to seize their land, overthrow their leaders, and pervert their way of life. The terms for this rhetorical assault on Islam included: The "War" as a "Crusade,"[28] and terrorist activity as "Jihad."[29] Moreover, the 'fighting words' that President Bush used to unify his base supporters and to persuade Americans at home of the need and the appropriateness of war as a response—"Bring 'em on!,"[30] "Dead or Alive,"[31] "Mission Accomplished"[32] and "We're kicking ass!"[33]—reinforced to the rest of the world the President's "cowboy" image and largely unilateral approach to foreign policy.

The adoption of relatively clear and simple language to describe a complex, fluid and emergent problem, while rhetorically satisfying in the days and weeks after 9/11, has increasingly stifled our ability to turn the tide in the global "war" of ideas.

Failed Dialogue, Failed Diplomacy

One of the unanticipated consequences of President Bush's rhetoric was its increasing tendency to overshadow diplomatic efforts at bolstering support for the U. S. and its policies abroad. From September of 2001 until October 2005, the Pew Center's reports on the image of the U. S. fell dramatically, achieving by the Summer of 2007 the lowest recorded rating in our history.[34] This lack of support for the U. S. and its policies carried over into surely well-intentioned, but flawed diplomatic efforts designed to re-establish America's positive standing in the world. In an attempt to restore confidence in the U. S., President Bush turned to his long-time campaign advisor, Karen Hughes, to use her considerable and time-tested U. S. political marketing talents to reverse the downward spiral, particularly in the Middle East. She was appointed by Condoleezza Rice as Under Secretary of State for Public Diplomacy on March 14, 2005.

Karen Hughes' approach to influence and campaign communication appears to be one she inherited from Charlotte Beers, former Under Secretary of State for Public Diplomacy and Public Affairs. She was appointed after 9/11 and charged with developing a campaign to counter anti-American sentiment, particularly in the Middle East. Charlotte Beers made her reputation as a Madison Avenue advertising whiz. She was the chairwoman of two of the top 10 advertising agencies in the country, the first woman Product Manager for Uncle Ben's Rice, and named one of the most powerful women in America by *Fortune* magazine.[35] Her approach to public diplomacy was, not surprisingly informed by a public relations and advertising perspective. Indeed, she advocated that paid media campaigns were the key to countering anti-American sentiment. In those campaigns, she was interested, primarily, in sending the right unified message. In a speech at the Washington Institute for Near East Policy in 2002, Ms. Beers remarked that one of the "issues that interests us greatly at Public Diplomacy at the State Department today . . [is] What's the message? Who's the messenger? And where can it be delivered and on what timeline?" Her communication goals were also largely sender and message centered, to wit :

- The first is to inform our many publics of the content of U.S. policy—accurately, clearly and swiftly;
- Next, re-present the values and beliefs of the people of America, which inform our policies and practices;
- Third, define and provide dimension to the role that democracy plays in engendering prosperity, stability, and opportunity; and
- Fourth, communicate our concern for and support of education for the younger generations.[36]

When Karen Hughes took over the role of Under Secretary, she brought with her experience as a journalist, and more importantly, a political campaigner. Political campaigns operate with assumptions borrowed from advertising and public relations. So, while the players change, the communication theory stays the same. Once again, while these public relations and advertising models work well in the affluent, consumer-driven commodity market of the West, in the War on Terrorism they reduce ideas such as freedom, democracy, and religious pluralism as if they were no different than one's "choice" of competing soft drink or insurance companies.

This troublesome flaw in communication strategy was compounded by a corresponding failure to use language and context to effectively and persuasively represent an understanding of the complexities involved in Middle Eastern gender and religion politics. In what was billed as a "listening tour" to learn more about the culture and customs of the Middle East, Karen Hughes used her trip as another opportunity to deliver the Bush message of freedom, democracy and equality.[37] This strategy was not warmly received by the host audiences. Women rebuffed the implication that their lives were less rich and fulfilling than those of their American counterparts and challenged the claim that the U.S. was a "friend." What good is freedom without God? What good is democracy without obedience to God? And what good is equality when the price of equality is a disruption of the roles of women and women as defined by God? And why are you here speaking to us when you should be at home with your children? Hughes' responses were largely ones of shock and surprise, and she retreated from engagement to the known safety of simplistic campaign assertions and a seemingly defensive posture. What had been a difficult, but hopeful opportunity for dialogue became, for all the world to see, another instance of American failed diplomacy, hypocrisy, and the arrogance of power: It was a failure not because the key players were not sincere or committed to making progress, but because it was informed by the uncritical and unreflexive adoption of a bad communication theory.

Hughes was not alone in this. Condoleezza Rice became Secretary of State in 2005. She brought with her to the position an approach to communication that was born of technical proficiency with information and tempered by the Cold War; She was a Soviet expert who served as a national security advisor. George Bush could not have found a person more in tune with an "us versus them" attitude and penchant for power politics that emphasized military strength over rhetorical finesse. As Joseph Nye has argued, the role of diplomacy is the marshalling of "soft power,"[38] of the power of building coalitions through dialogue and reasoned influence. Yet Secretary Rice, much like her boss, eschews soft power in favor of the Cold War style of hard-line posturing that she learned from her stylistic mentor, Henry Kissinger.[39] One characteristic of that rhetorical style is the staunch unwillingness to legitimize any opposition by talking to, or carefully listening to, their leaders.

For example, when the President of Iran sent an open letter to the President, the Secretary of State refused to respond officially, claiming that it did not come through approved diplomatic channels and therefore did not count as an authentic form of communication. One of the consistent drawbacks of the Bush administration communication policy, again born of a reliance on a linear model of communication, has been an unwillingness to recognize that, in a globally mediated environment, there are always several channels operating simultaneously and that we can no longer afford to recognize, respond to, and privilege only the "official" ones.

While U. S. political opinion may be shaped by CNN, Fox and MSNBC, it is increasingly influenced by "unofficial" and alternative news sources, such as *The Daily Show*, the *Daily Kos*, Rush Limbaugh and a multitude of other partisan grassroots blogs and websites. So it is should be no surprise that global political opinion is not informed solely or even primarily by the State Department's messages delivered through official channels. Perhaps more importantly, it is shaped by non-traditional news channels and new media resources.

The result is a set of new communication challenges that cannot be resolved by relying upon traditional media and diplomatic lines of communication for influence worldwide. This is a post-Cold War information technology lesson that has yet to be embraced by current administration strategy.

This lack of understanding of the new global communication context, was revealed poignantly in the controversy surrounding the Iranian President's letter to President Bush (see Chapter 5 of this volume). The letter not only gained media attention, but as a result of our silence on the issues he raised, President Ahmadinejad reinforced the global perception that the U.S., long considered the champion of free speech and dialogue, was inexplicably unwilling to engage a legitimate leader on issues of mutual concern. Whether or not the letter represented a challenge to traditional diplomatic channels did not matter; what did matter was that it was *interpreted* to mean that our values could be challenged and we would remain silent, therefore weakening our posture on the world political stage, particularly in those outlets already predisposed to question our integrity.[40]

Cartoon depicts President Bush's unwillingness to engage. By permission of Dwane Powell and Creators Syndicate, Inc.

This strategy of taunting the U. S. was repeated only months later by Venezuela's President Hugo Chavez (Stout, 2006)[41], and once again, the official diplomatic response was to denigrate the integrity of the speaker and remain silent on the issues. Despite the fact that many observers found Chavez's rhetoric to be roughly the equivalent of how Mussolini parroted Hitler prior to WWII,[42] by failing to respond to his charges and invitations to address the issues, the U. S. fared less well. We were characterized as having more in common with a schoolyard bully than the leading champion of free speech and the open expression of ideas.[43]

This lack of a good defense against the rhetoric of tyrants on the world's stage was negatively complemented by a corresponding lack of a good offense to reassure our friends and allies. There appeared to be no coherent narrative tying the United States' policies and actions in the War on Terror to larger issues of a moral compass informed by the consistent application of core human values. This was nowhere more apparent than in the administration's missteps in their handling of Abu Ghraib,[44] and Guantanamo Bay, Blackwater[45] (and, closer to home, the Walter Reed Army Medical Center neglect[46]) scandals. As Philip Carter, writing in *The Washington Monthly* put it:

> *A generation from now, historians may look back to April 28, 2004, as the day the United States lost the war in Iraq. On that date, "CBS News" broadcast the first ugly photographs of abuses by American soldiers at Baghdad's Abu Ghraib prison. There were images of a man standing hooded on a box with wires attached to his hands; of guards leering as they forced naked men to simulate sexual acts; of a man led around on a leash by a female soldier; of a dead Iraqi detainee, packed in ice; and more. The pictures had been taken the previous fall by U.S. Army military police soldiers assigned to the prison, but had made it into the hands of Army criminal investigators only months later, when a soldier named Joseph Darby anonymously passed them a CD-ROM full of prison photos. The images aroused worldwide indignation, and illustrated in graphic detail both the lengths to which the United States would go to get intelligence, and the extent to which those efforts had been corrupted by the exigencies of the difficult war in Iraq.*
>
> *Two days later,* The New Yorker *published a report on Abu Ghraib by Seymour Hersh. Hersh won a Pulitzer Prize in 1970 for his reporting on the U.S. Army's atrocities in Vietnam; now he had come full circle, documenting the full extent of the abuses at Abu Ghraib and the Army's initial efforts to investigate them. Hersh's reporting—which forms the nucleus of his new book,* Chain of Command*—helped launch nearly a dozen different criminal investigations into what former vice president Al Gore dubbed "the American Gulag," the extraterritorial chain of prisons and detainment centers, stretching from Guantanamo Bay to Afghanistan, set up by the Bush administration to hold suspected terrorists.*[47]

How could a nation devoted to justice, liberty, the fair and humane treatment of prisoners of war and the legal protections guaranteed by our Bill of Rights sanction the mistreatment, illegal detainment, and abuse of foreign citizens? In terms of world opinion, our arrogance was only matched by our hypocrisy.

Once again, the need for open and honest communication on the issues, for an ability to admit wrongdoing, and for taking responsibility for our actions was curiously absent in the administration's stonewalling response, replete with denial, legal maneuvering, and outright obfuscation. "Any short term gains achieved by such strategies merely serve to weaken our institution[s] in the long run" (Chiarelli & Smith, 2007, p. 12). These strategic missteps in the arena of public diplomacy have contributed to the erosion of the image and credibility of America at home and abroad.

Reversing the failures of diplomacy and counterterrorism cannot happen by openness and honesty alone. It will take a coordinated, smart and complex two-pronged approach to a campaign to boost our image and reputation and, simultaneously, discredit and disrupt the enemies' campaigns of politics and violence leveled against the West. Our campaign should be measured by realistic expectations, benchmarks, and timetables. This should not be hard to do. As Thomas Friedman pointed out in an op-ed for the *New York Times*: "How could an administration that was so good at Swift-boating its political opponents at home be so inept at Swift-boating its geopolitical opponents abroad?"[48] By failing to effectively counter the communication campaign against us, Bush is effectively "losing a P. R. war to a mass murderer."

Beating Us at Our Own Game: Failure to Counter Terrorists' Propaganda

One surprising feature of terrorist networks is their sophisticated use of media, especially new media, to reinforce their message and to recruit new zealots (see Corman & Schiefelbein, this volume). Media scholars consistently point out the fact of a media elite that dominates Western outlets, but this argument should be augmented by a realization that our enemies have appropriated the tactics of successful branding and advertising strategies that dominate political campaigns. As Robert McChesney notes "the internet is being colonized by the corporate landscape. The elite media outlets have gradually replaced the pursuit of truth with the undemocratic placement of planted publicity stories"[49] but scholars throughout the world have been slow to pick up on the fact that the media "elite" that is planting publicity stories in the Middle East,

Europe, and Asia includes terrorist organizations and their messages of hate and violence.

Once again, for students of political rhetoric, the strategies violent extremists deploy are not new.[50] Indeed, they look remarkably familiar:

- Black and white fallacy
- Big lie
- Demonizing the enemy
- Disinformation
- Flag-waving
- Oversimplification
- Repetition
- Scapegoating
- Slogans
- Stereotyping
- Testimonial
- Virtue words

One reason the U.S. has not been able to effectively counter terrorist propaganda efforts is that despite using the same proven techniques, we do not have a credible message, a credible messenger(s), we have not adequately organized or engaged local resources to the task, and we rarely admit mistakes even when they are fairly obvious to everyone in the world.[51]

Terrorist propaganda cannot be countered with appeals to reasoning or truth that fall outside of the target audiences' accepted cultural understandings and beliefs. Nor can success be accomplished simply by denying the validity of terrorists' messages. Nor can we allow "fixed meanings" already reinforced routinely in mosques and over the airwaves to "stand" unchallenged by responding to propaganda with like strategies (see Corman, Trethewey & Goodall, this volume). What we need is a strategy to interrupt and disrupt the interpretive frameworks that exist for audiences around the globe, including, but not limited, to violent extremists. Rhetoric that does little more than say, "We're right and you're wrong" does not provide any resources for audiences on either side to interpret meanings anew. Instead, propaganda works only to reinforce the "base" of true believers.

One obvious place where we have failed to disrupt a "fixed meaning" as it currently exists in some interpretive communities is in the concept of "martyrdom" that is used to recruit young idealists to the cause and that is used to encourage them to give up their lives (as well as the lives of innocent others) for the cause. Our strategy has been to try to discredit "suicide bombers" as being counter to the teachings of the Koran, often relying on "the mythology of martyrdom" to scapegoat the practices associated with it.[52]

But this English language message strategy does not always translate effectively into Arabic, where there is no literal or cultural equivalent for the term "suicide bomber" and where the local culture and religion treat such acts as "martyrdom."[53] Another example is in how successful terrorist leaders—like bin Laden—have been in framing, rapidly and repeatedly, the insurgency in Iraq as a noble resistance movement. They take credit for all insurgent groups and all insurgent causes despite the obvious fact that many insurgents see al-Qaeda as the enemy and most of the insurgent operations in Iraq are more a product of tribal civil war than of raw opposition to the U. S. forces.[54]

Lt. General Peter Chiarelli and Major Stephen Smith, both of whom have served multiple tours in Iraq, argue that the "most decisive factor that will determine who emerges victorious in current and future wars is which side can gain consistent advantage in the holistic information environment that plays out across the globe, near and far from the 'front lines' (2007, p. 10). Importantly, for communication scholars and practitioners, these military specialists argue, as we do, that "currently, we do not respond well enough to deal effectively with enemies who can say whatever they want without retribution" (p. 10).

Clearly, communication scholars are among those who are poised to develop anti-fundamentalist, rapid, and proactive responses to the often video-camera wielding violent extremists' messages of violence on blogs, websites and other mediated fora, whether it is in the form of exalting "martyrs" or encouraging fundamentalist communication, or (falsely) discrediting agents of the U. S and their allies.[55]

Shock and Awesome Silences: The Failure to Use the Power of Victims' Narratives and Everyday Lives

What is called for is a new, theoretically informed guide to communication practices that are grounded in the cultural and religious pragmatics of everyday life. These counter-fundamentalist pragmatics make creative use of *humility and ambiguity* as well as of alternative message sources and complexities to change the dialogue, disrupt the extant conversation, and to create opportunities for evoking alternative meanings among local populations.[56] To change the nature of the talk requires more than understanding the violent extremists' communication strategies; it requires rethinking and redeploying what counts as "communication" in their, and in many ways our, worlds.

It also requires making use of what we know about appeals to the targeted age group (15—30), and a demonstrated willingness to abandon a message-based communication strategy that no longer works. Finally, it requires an ability to work in ambiguous and less controlled environments and emergent networks and to embrace empowering narratives as key to change, not repetitive and sloganistic messages as the core component of our longer-term strategy for improving our image while discrediting theirs. As Eisenberg points out, "contrary to fundamentalist thinking, the moral high ground rests on a foundation of ambiguity and humility" (2007, p. 296).

The future of strategic communication will likely center on narratives, particularly as those narratives emerge from, inform, and transform networks[57] of "netizens." [58] This narrative approach to the war of ideas must include a clear recognition that change on the street means targeting people where they live. Ordinary citizens that we want to reach spend their free time much as we do: reading books and newspapers, surfing the net, watching television and playing video games, but not necessarily watching the news, including *al-Jazeera.* What influences audiences worldwide is what attracts them and keeps their attention, increases their visual pleasure, their aural enjoyment, and their social cache in the own communities. Thus, if we want to change the local conversation, we must find ways to engage audiences where they are using old and new information technologies and find more creative approaches to what counts as a truly persuasive "communication event."

For example, one by-product of our knowledge of violent extremist media strategies has been the realization that we need to engage them in hyperspace and using local resources already in their communities perhaps more than we need to rely on traditional face-to-face encounters by diplomatic and military elites.[59] After all, how—and when—do you speak to someone who hates you? Chances are, you cannot or you do not. But *if* you can create surrogates, such as avatars in virtual worlds or characters in locally produced sit-coms or bloggers who can successfully engage issues without being directly tied to U. S. government sources, then perhaps there is a mediated alternative to traditional communication channels and diplomatic strategies.

We were pleased to learn that that State Department has recently created a Digital Outreach Team that includes Arabic-speaking diplomats who are posting to mainstream websites in an effort to "take a more casual, varied approach to improving America's image in the Muslim world."[60] Brent E. Blaschke, the project director, said the "idea was to reach 'swing voters,' whom he described as the silent majority of Muslims who might sympathize with *Al Qaeda* yet be open to information about United States government policy and American values" (MacFarquhar, 2007).

Again, encouragingly, the State Department has, in this instance, moved away from a traditional sender-oriented model of communication.[61] "We try to put ourselves in the mindset of someone receiving the message," said Duncan MacInnes, the director of the Counterterrorism Communication Center, of which the Digital Outreach Team is one branch. "Freedom for an Arab doesn't necessarily have the same meaning it has for an American. Honor does. So we might say terrorism is dishonorable, which resonates more." While we think this practice is a positive step in the right direction, we will continue to encourage U. S. strategic communication practitioners to think even more creatively about reaching, developing and listening to new audiences' narratives, particularly victims' narratives.

It will also be useful to make more productive and poignant use of local narratives, released by local producers for local audiences, that describe the lives of victims and survivors of terrorist attacks and suicide bombers.[62] For audiences accustomed to only hearing about "martyrdom," and who operate in Arabic languages wherein there is no literal translation of "suicide bomber" at all, hearing and seeing the devastating personal and familial effects of so-called "martyrdom" may alter the conversation in ways that direct public relations campaigns staged by the military and diplomatic corps cannot.

By empowering victims' voices, by encouraging the telling of stories that resonate with local populations, the landscape for terrorist rhetoric becomes far more complicated than is now the case. In the end, making terrorists and their leaders locally responsible to citizens who want safe and secure neighborhoods and better lives for their children and grandchildren should prove a far better communication strategy than simply repeating stale, failed messages about democracy and freedom.

Conclusions

Words are weapons in the struggle against violent extremism. Choices about who should deliver them; how, when, and where they should be delivered, and contingency plans for dealing with likely fallout, are all part and parcel of a sound strategic communication plan. To date, the United States has yet to articulate a strategic communication policy that incorporates state-of the-art communication theory about the selection and use of these "weapons."

Weapons of mass persuasion we have deployed against civilian populations have proven destructive. And, just as abandoned weapons on a battlefield often lead to unintended casualties, words in the struggle against violent extremism echoing through cyberspace have led to equally dangerous unintended consequences: the recruitment of a new generation of violent extremists who use our words against us. This volume represents our efforts to imagine a new approach to strategic communication that can help to move the public, mediated, and even interpersonal conversations away from fundamentalist forms of communication to those that have the power to open us to new interpretive possibilities, including the de-escalation of violence, and the growth of social movements, that may be largely mediated, toward peace, human rights, and a globally engaged citizenry/netizenry.

Notes

1. For a history of the use of this term, please see *http://en.wikipedia.org/wiki/Strategery*
2. The term "war of ideas" is used here because it is in wide circulation (see, for example, Walid Phares, *The War of Ideas*). However, the term is highly problematic for us for reasons made clear in this chapter as well as in Chapters 7 and 8. For most of the rest of the book, we will use alternative constructions such as the "struggle against violent extremism" except when quoting directly from sources who use the term "war" or in reference to the "Global War of Ideas" as a concept put forward by the Bush administration.
3. *http://www.guardian.co.uk/commentisfree/story/0,,2107801,00.html*
4. Pew Global Attitudes Project (2006, June 13). http://pewglobal.org/reports/display.php?PageID=825
5. Senator Thomas Kean, chairman independent commission investigating the 9/11 terror attacks, claimed that the attacks could and should have been prevented. See http://www.cbsnews.com/stories/2003/12/17/eveningnews/main589137.shtml
6. For a timeline of critical communication events that have led to the decline of our status and image in the world, see Rich (2006).
7. See H. L. Goodall, Jr., "Why We Must Win the War on Terror: Communication, Narrative, and the Future of National Security," *Qualitative Inquiry, 12,* (February, 2006), 30–59. For an historical summary of the "progress" of communication theories, see Eric M. Eisenberg, H. L. Goodall, Jr., and Angela Trethewey, *Organizational Communication: Balancing Creativity and Constraint*, 5th ed. New York:

Bedford/St. Martin's, 2007, chapter two.

8. Burke, K. (1962). *A Rhetoric of Motives*. Berkeley, CA: University of California Press.
9. For explication of the culture of fear, see Altheide, D. L. (2006). *Terrorism and the Politics of Fear*. Walnut Creek, CA: Altamira Press; see also his *Creating Fear: News and the Construction of Crisis.* Hawthorne, NY: Aldine De Gruyter, 2002.
10. Giroux, H. A. (2004). "Beyond belief: Religious fundamentalism and cultural politics in the age of George W. Bush." *Cultural Studies—Critical Methodologies 4,* 415-426.
11. See Anderson, J. A. (1996). *Communication Theory: Epistemological Foundations.* The Guilford Press; see also Weick, K. (1979), *The Social Psychology of Organizing.* New York: Random House and Berger, P., and Luckmann, T. (1966), *The Social Construction of Reality.* New York: Anchor Books.
12. Fisher, W. (1987). *Human Communication and Narration: Toward a Philosophy of Reason, Value, and Action.* Columbia, SC: University of South Carolina Press; see also Mumby, D. K. (1993). *Narrative and Social Control: Critical Perspectives.* Walnut Creek, CA: Sage. For philosophical underpinnings, see Rorty, R. (1989), *Contingency, Irony, and Solidarity.* Cambridge: Cambridge University Press.
13. See Eric M. Eisenberg, *Strategic Ambiguities: Essays on Communication, Organization, and Identity.* Thousand Oaks, CA: Sage, 2006. See also William B. Hart II and Fran Hassencahl, "Culture as Persuasion: Metaphor as Weapon," in L. Artz and Y. R. Kamalipour (eds.), *Bring 'em On: Media and Politics in the Iraq War.* Lanham, MD: Rowman & Littlefield, 2005, pp. 85–100.
14. "Sensemaking" is a Weickian term; see Weick, K. (1995), *Sensemaking in Organizations.* Thousand Oaks, CA: Sage.
15. See H. L. Goodall, Jr., "Mysteries of the Future Told: Communication as the Material Manifestation of Spirituality," *World Communication Journal 22* (December, 1993), 40–49.
16. See Eric M. Eisenberg, H. L. Goodall, Jr., & Angela Trethewey, *Organizational Communication: Balancing Creativity and Constraint,* 5th ed. New York: Bedford/ St. Martin's Press, 2007.
17. See Coleman, Peter T. "Intractable Conflict." Morton Deutsch and Peter T. Coleman, eds., *The Handbook of Conflict Resolution: Theory and Practice.* San Francisco: Jossey-Bass Publishers, 2000, pp. 428–450.
18. The British officially dropped the use of the term on April 17, 2007. See http://www.thestar.com/News/article/203803 ; the American Democratic Party banned the use of the term in the Pentagon budget proposal for 2007; see *http://www.telegraph.co.uk/news/main.jhtml?xml=/news/2007/04/10/wusdems10.xml*
19. For further explication of this point, see H. L. Goodall, Jr., "Fieldnotes from Our War Zone: Living in America During the Aftermath of September Eleventh," *Qualitative Inquiry 8* (April 2002), 74–89.

20. It is worth noting that even President Bush's most trusted advisors, including Donald Rumsfeld and Stephen Hadley, urged him to abandon the use of the term "war on terror" in favor of "global struggle against violent extremism" as late as August 2005. See *http://www.telegraph.co.uk/news/main.jhtml?xml=/news/2005/08/05/wterr05.xml* .
21. Similar debates have emerged on the appropriate vocabulary for naming terrorists and their activities. While President Bush and his colleagues invoked the language of war immediately after 9/11, that language was not the only possibility before him. Indeed, in October of 2001, the Congressional Research Service reported that the question of "whether to treat the attacks as acts of war or criminal acts has not been fully settled" (Elsea, 2001, p. 1). While the language and practice of treating terrorists as soldiers and enemy combatants have emerged as victor in this rhetorical struggle, the debate still continues. Recently, Former Supreme NATO Commander Wesley Clark and Professor Kal Raustiala (2007, August 8) argued that terrorists should be treated and tried as criminals, not combatants. Defining terrorists as "combatants" is a mistake, they claim, because it gives terrorists the status of soldiers, elevates the cause of terrorism, endangers American political traditions, and erodes our credibility abroad.
22. On September 11 and afterward, those chief advisors included Vice President Dick Cheney, National Security Advisor Condoleezza Rice, CIA Director George Tennet, Secretary of State Colin Powell, political advisors Karl Rove and Karen Hughes, and White House spokesperson Ari Fleisher. See CBS interview with President Bush at *http://www.cbsnews.com/stories/2002/09/11/60II/main521718.shtml* . See also Bob Woodward, *Bush at War*. New York: Simon & Schuster, 2002.
23. For a discussion of this rhetorical strategy, see H. L. Goodall, Jr., "Why We Must Win the War on Terror: Communication, Narrative, and the Future of National Security," *Qualitative Inquiry, 12,* (February, 2006), 30–59.
24. See Walid Phares, *The War of Ideas: Jihadism Against Democracy*. New York: Palgrave/Macmillan, 2007.
25. The unfortunate use of the term "crusade" was widely disparaged in the European and Middle Eastern press. See *http://www.csmonitor.com/2001/0919/p12s2-woeu.html*
26. Michael Scheuer, *Imperial Hubris: Why the West Is Losing the War on Terror*. Washington, D.C.: Potomac Press, 2004.
27. *The 9/11 Commission Report: Final Report of the National Commission on Terrorist Attacks Upon the United States*. New York: W. W. Norton, 2004.
28. http://archives.cnn.com/2001/US/09/20/gen.bush.transcript
29. *http://www.opinionjournal.com/extra/?id=95001224*. The term "jihadist" as a negative connotation and rhetorical equivalent to terrorist has sparked serious attention among scholars, diplomats, and politicians. Western political analysts and

politicians who are familiar enough with Islam recognize that "jihad" is a spiritual obligation of a believer, but the broader scholarly point is that "the new proposition advanced by scholars in the West that a non-violent, inner, and personal jihad is the 'real one' can be tested only in the wake of a cultural, widely accepted principle that the historical theologically endorsed jihad warfare is over, and not just suspended or hidden" (Phares, 2006, p. 2004). Hence, the term is contested even by those who understand its deep historical and cultural origins. But narratively and rhetorically, by granting to the terrorists the holy terrain of jihadist, we play, once again, into the hands of the violent extremist who used our characterization to foment anti-American sentiment and to recruit the next generation. If opposition to U. S. values needed additional rhetorical fuel for an already blazing narrative fire, these fighting words provided it. It also led to an increase in the number of recruits and splinter groups.

30. *http://www.usatoday.com/news/world/iraq/2003–07–02-bush-iraq-troops_x.htm*
31. http://archives.cnn.com/2001/US/09/17/bush.powell.terrorism
32. http://www.cnn.com/2003/ALLPOLITICS/05/01/bush.carrier.landing
33. *http://select.nytimes.com/2007/09/09/opinion/09rich.html?_r=1&adxnnl=1&oref=slogin&adxnnlx=1190396261-TnXY3cj9J6CA1Vsymveh6g*
34. See Andrew Kohut's analysis of the first two years following 9/11 (*http://people-press.org/commentary/display.php3?AnalysisID=63*) and the later comprehensive listing of all surveys (*http://www.publicdiplomacy.org/14.htm*).
35. For a fuller biography for Ms. Beers, please see the State Department's website at http://www.state.gov/outofdate/bios/b/5319.htm
36. For a complete transcript of this speech entitled, "U. S. Public Diplomacy in the Arab and Muslim Worlds," please see *http://www.state.gov/r/us/10424.htm*
37. See the account published in *The Washington Post* for a perspective on why campaign tactics didn't work on the carefully selected audiences she addressed: *http://www.washingtonpost.com/wp-dyn/content/article/2005/09/29/AR2005092901290.html*
38. Joseph Nye, *Soft Power: The Means to Success in World Politics*. Washington, D.C.: Public Affairs Press, 2004.
39. Marcus Mabry, *Twice as Good: Condoleezza Rice and Her Path to Power*. Rodale, 2007.
40. Cartoon taken from the political blog TruthDig: *http://www.truthdig.com/cartoon/item/20061129_powell_grade_school_diplomacy/*
41. For a fuller analysis of Chavez's rhetorical strategy, see Morris's (2007, March 27), "Sanctioning the Devil" at the Consortium for Strategic Communication's blog: http://comops.org/journal/2007/03/27/sanctioning-the-devil/
42. See, for example, *http://www.zmag.org/content/showarticle.cfm?ItemID=11022*
43. See, for example, *http://www.slate.com/id/2079678/*
44. For a book-length treatment of the Abu Ghraib scandal as a broad, systems-level or "contaminated orchard" problem rather than a case of "rotten apples," see Mestrovic (2007).

45. *http://www.nytimes.com/2007/09/28/world/middleeast/28contractors.html?_r=1&ref=middleeast&oref=slogin*. See also the Blackwater website *http://www.blackwaterusa.com/*
46. See for example, Priest & Hull (2007).
47. *http://www.washingtonmonthly.com/features/2004/0411.carter.html*
48. *http://select.nytimes.com/2007/08/26/opinion/26friedman.html?hp*
49. cited in Snow, N. (2003). *Information war: American propaganda, free speech and opinion control since 9/11*. New York: Seven Stories Press, p. 30.
50. See the re-release of the classic propaganda text, Bernays, E., and Miller, M. C. (2007). *Propaganda*. Brooklyn, NY: Ig Publishing. For extended discussion of specific techniques, see Jowett, G. S., & O'Donnell, V. (2005), *Propaganda and persuasion*. Thousand Oaks, CA: Sage.
51. See Rich, F. (2007). *The greatest story ever* sold*: The decline and fall of truth in Bush's America.* London: Penguin.
52. See Hafez, M. M. (2007). *Suicide bombers in Iraq: The strategy and ideology of martyrdom.* Washington, D.C.: United States Institute of Peace Press, p. 230.
53. This idea for the problem of translation emerged in a seminar devoted to "Discrediting Suicide Bombers," CENTCOM/SOCOM and the University of Texas at San Antonio, San Antonio, TX, August 2007.
54. See Hafez, cited above, p. 89–90.
55. Chiarelli and Smith (2007) provide a poignant example of the information (read: interpretation) environment in which the military, state department, and civilian population are now operating. In early 2006, coalition forces killed 17 insurgents in Baghdad. The enemy then dragged the bodies of the dead to a nearby prayer room and staged the scene to make it look as though the insurgents had been executed. While the coalition was able to disconfirm the insurgents' framing of the story, the "massacre" story was the one that circulated widely, both in the streets and in the media, even after a hostage freed in the initial raid provided testimony supporting the coalition's rendering of events. Their larger point, however, is that events like this happen every day in Iraq, and the U. S. has "failed to take initiative or even effectively defend ourselves in the information environment" (p. 10). Oftentimes, the first story is the one that "sticks" in a media environment, even when it is not true. Thus, our need to develop rapid, proactive responses, and as John Rendon (personal communication) suggests, simultaneous methods for slowing down the enemies' communication cycles.
56. See Corman, Trethewey & Goodall in this volume.
57. Noted strategic communication expert and practitioner, John Rendon of The Rendon group made this prediction in a talk entitled "Strategic Communication" on September 10, 2007 in the Hugh Downs School of Human Communication at Arizona State University, Tempe, AZ.
58. This terms was first coined by Hauben and Hauben (1997) in their text, *Neti-*

zens: On the history and impact of usenet and the internet. More recently, Wikipedia defined netizens as those who "use the Internet to engage in activities of the extended *social groups* of the internetworks (i.e., giving and receiving *viewpoints*, furnishing *information*, fostering the Internet as an intellectual and a social resource, and making choices for the self-assembled communities)" (http://en.wikipedia.org/wiki/Netizen).

59. Fouts, J. He is editor of the forthcoming, "Politics and Play" (Peter Lang). He is on the editorial boards of *Games and Culture: A Journal of Interactive Media* (Sage), and *Place Branding* (Palgrave Macmillan)
60. Burgess, 2007 September 22.
61. After several circulating several white papers, delivering briefings to the State Department, and attending conferences around the world advocating this position, those of us in the Consortium for Strategic Communication, are heartened that our "message" may be finding traction among some key audiences.
62. For example, Brooke Goldstein's documentary, "The Making of Martyr" explores, among other things, how families of suicide bombers are impacted by the mythology of martyrdom and its material consequences. To download, go to *http://movies.aol.com/truestories/making-of-a-martyr*

References

Ananthaswamy, A. (2007). A life less ordinary offers more than just escapism. *New Scientist (Australia/New Zealand), 195*, 40.

Burgess, J. F. (2007, September 22). State Department discovers blogs. *Crossroad Arabia*. Retrieved September 23, 2007 from http://xrdarabia.org/2007/09/22/state-department-discovers-blogs

Burke, J. (2006, October 10). British stops talk of 'war on terror'. *The Guardian Unlimited. l http://observer.guardian.co.uk/politics/story/0,,1968668,00.html*

Chiarelli, P., & Smith, S. (2007, September-October). Learning from our modern wars: The imperatives of preparing for a dangerous future. *Military Review.* Retrieved September 23, 2007 from *http://usacac.army.mil/CAC/milreview/English/SepOct07/chialleriengseptoct07.pdf.*

Clark, W. K., & Raustiala, K. (2007, August 8). Why terrorists aren't soldiers. Retrieved September 18, 2007 from *http://select.nytimes.com/gst/abstract.html?res=F10910F935540C7B8CDDA10894DF404482*

Eisengberg. E. (2007). *Strategic ambiguities: Essay on communication, organization, and identity.* Thousand Oaks, CA: Sage.

Elsea, J. (October 29, 2001). Trying terrorists as war criminals. CRS Report for Congress. Retrieved September 19, 2007 from *http://fpc.state.gov/documents/organization/6270.pdf*

Hauben, M., & Hauben, R. (1997). *Netizen: On the history and impact of usenet and the internet*. Hoboken, NJ: Wiley—IEEE Computer Society Press.

MacFarquhar, N. (2007, September 22). At State Dept., blog team joins Muslim debate. *New York Times*. Retrieved September 23, 2007 from http://www.nytimes.com/2007/09/22/washington/22bloggers.html.

Mestrovic, S. G (2007). *The trials of Abu Ghraib: An expert witness account of shame and honor*. Boulder, CO: Paradigm Publishers.

Priest, D., & Hull, A. (2007, February 18). Soldiers face neglect, frustration at army's top medical facility. *The Washington Post*. Retrieved August 25, 2007 from *http://www.nytimes.com/2007/09/22/washington/22bloggers.html?th=&adxnnl=1&emc=th&adxnnlx=1190585257-PyyXSH58NlHgOlOcQttllQ*

Schmitt, E., & Shanker, T. (2005, July 26). U.S. officials retool slogan for terror war. Retrieved September 21, 2007 from http://www.washingtonpost.com/wp-dyn/content/article/2007/02/17/AR2007021701172.html *http://www.nytimes.com/2005/07/26/politics/26strategy.html?pagewanted=all*

Stephenson, R. (2005, August 4). President Makes It Clear: Phrase Is 'War on Terror.' Retrieved September 21, 2007 from *http://www.nytimes.com/2005/08/04/politics/04bush.html?ex=1280808000&en=b191eee9455d238d&ei=5088*

Stout, D. (2006, September 20). Chavez calls Bush 'the Devil' in U. N. Speech. *New York Times*. Retrieved September 1, 2007 from http://wSe B20/world/americas/20cnd-chavez.html?ex=1158984000&en=aa643a6bdfaf2377&ei=5087%0A

Ward, O. (2007, April 17). British drop use of "war on terror." Retrieved September 21 from http://www.thestar.com/News/article/203803

CHAPTER TWO

Strategic Ambiguity, Communication, and Public Diplomacy in an Uncertain World

Principles and Practices

H.L. GOODALL, JR., ANGELA TRETHEWEY, AND KELLY MCDONALD

Introduction

There is widespread recognition that the U. S. public diplomacy efforts worldwide have failed. In response to this image crisis, the Pentagon, State Department, and other agencies of the federal government are currently seeking new models for message strategy, coordination, and alignment.

There are two major reasons for failures of communication in public diplomacy: (1) reliance on an outdated one-way model of influence, and (2) an inability to prepare for, or respond to the jihadi media and message strategy that has thus far dominated local cultural interpretations of U.S. diplomatic objectives.

These failures can be addressed if the U.S. recognizes the need for a new way of thinking about *ambiguity as strategy* in strategic communication initiatives. Strategic ambiguity recognizes that a powerful vision for change among diverse constituents requires an ability to empower local interpretations of its meaning in order to build relationships to that vision without insisting on a fixed meaning for it or alienating potential allies because of it. Ambiguous but mindful communication practices are required in uncertain times, particularly when dealing with audiences we neither fully understand nor trust.

Five principles to guide strategic communication policy are: (1) practice strategic engagement, not global salesmanship; (2) do not repeat the same message in the same channels with the same spokesperson and expect new or different results; (3) do not seek to control a message's meaning in culures we do not fully understand; (4) understand that message clarity and perception of meaning is a function of relationships, not strictly a funcion of word usage; and (5) seek "unified diversity" based on global coopeation instead of "focused wrongness"based on sheer dominance and power.

Background

In a report issued on June 13, 2006 by the Pew Global Attitudes Project, the first paragraph offers a devastating analysis of current world attitudes toward the U.S. It begins:

> *America's global image has again slipped and support for the war on terrorism has declined even among close U. S. allies . . . The war in Iraq is a continuing drag on opinions of the United States, not only in predominantly Muslim countries but in Europe and Asia as well. And despite growing concerns over Iran's nuclear ambitions, the U. S. presence in Iraq is cited at least as often as Iran—and in many countries much more often—as a danger to world peace (p. 1).*

During a visit to the Army War College in March, Rumsfeld said, "If I were grading, I would say we probably deserve a D or D plus as a country as to how well we're doing in the battle of ideas that's taking place in the world today. ... We have not found the formula as a country."

If this report is correct in documenting a continuous (since 2002) downturn in favorable attitudes toward the U.S. and its policies in the Middle East—and we believe it is—then the U.S. is losing the war of ideas.

The question is no longer "why?" The Pew report provides evidence that despite considerable expense, U.S. attempts to engage in strategic communication campaigns in the Middle East and around the world have failed, and that those failures have contributed significantly to an overall loss of credibility and prestige throughout the rest of the world. This critical assessment follows a published report about a Department of Defense plan for a comprehensive revision of its approach to public diplomacy ("The Roadmap," *U.S. News & World Report*, May 29, 2006) that will guide the work done in its new established Strategic Communication Center in Omaha, Nebraska.

Key to the Pentagon's new approach will be an emphasis on message analysis, coordination, and alignment among the various groups and agencies responsible for issuing statements, press releases, videos, news reports, and for making speeches. This emphasis on alignment is necessary because meanings attributed to messages are—in a global mediated environment—interpreted locally and rebroadcast within those locally interpreted frameworks to audiences who neither share our native language or culture and who are fundamentally dubious about the truth value of our messages.

The quiet admission of the U.S. failure to communicate a coherent and believable message and the Pentagon's investment in a new Strategic Communication Center are laudable steps in what promises to be a longer-term global communication campaign to regain our good standing both in the Muslim-dominated Middle East as well as on the world's multicultural, multiethnic, religiously and politically-diverse mediated stage.

The first step in the rehabilitation process should benefit from an understanding of why the existing communication strategy for public diplomacy in the Middle East failed. This chapter addresses that question by (1) providing a synthesis of academic studies that light on the two reasons for communication breakdowns in our "monologic" campaign to win the hearts and minds of Muslims as well as others, and the unlikely success of "dialogic" strategies; (2) offering a new "strategic ambiguity" model for strategic communication campaigns in the Middle East and elsewhere based on a middle-range theory that negotiates the communicative space between monologue and dialogue; and (3) providing five pragmatic principles of strategic ambiguity to guide the formulation of a new strategic communication in public diplomacy policy.

Two Reasons for Communication Failure

There are two major reasons for a failure of communication in public diplomacy: 1) reliance on an outdated one-way model of influence, and (2) an inability to prepare for, or respond to the jihadi media and message strategy that has thus far dominated local cultural interpretations of U.S. diplomatic objectives.

Reliance on a One-Way Model of Influence

For the past fifty years, the dominant U. S. approach to communicating with people living in regions of the world where we have strategic and economic interests has been informed by what communication theorists refer to as the "one-way model" (Dutta-Bergman, 2006; Goodall, 2006a).

This one-way model, derived from an early engineering study of telephone communication by Claude Shannon and Warren Weaver in the late 1940s (and refined slightly by public relations practitioners in the 1950s), posits that:

- *Messages* (verbal and nonverbal) move from a sender (or source), through a *channel* (e.g., airwaves, lightwaves), to a *receiver*;
- Meanings are contained in the words chosen by the sender and are *passively interpreted* by receivers (assuming the sender and receiver share the same language code and culture); and
- *Repetition* of the same message, sent through the same channels to the same receivers over time reduces outside interference with the intended message (i.e., noise) and improves the likelihood of achieving the intended interpretation and outcome (Goodall & Goodall, 2006).

No audience member is truly "passive" in her or his interpretation of a message.

The limitations of this model begin with the understanding that no audience member is truly "passive" in her or his interpretation of a message. The human mind *actively* engages words and actions within a particular context; places them into *pre-existing* historical, cultural, and political frameworks; and evaluates the meanings of the message based on *perceptions* of a source's credibility, intention, trustworthiness, and caring (Corman, Hess, and Justus, 2006; Peters, 1999; Hayakawa, 1978).

It is also true that meanings are *not* solely contained in the words chosen to convey a message. Human beings aren't dictionaries, but cultural interpreters of meanings. As the famous David Berlo dictum has it: "Meanings are in people, not in words" (Berlo, 1960). Berlo went on to say:

- People can have similar meanings only to the extent that they have had, or can anticipate having, similar experiences;
- Meanings are never fixed; as experience changes, so meanings change;

- No two people can have exactly the same meaning for anything.

These long established principles of meaning-making underscore the idea that a one-way transmission model has limited utility to organize public opinion in the realm of public diplomacy. If the U.S. government's strategic goal is to win the hearts and minds of diverse others, the most effective form of communication is *dialogue*, not monologue.

There is a wealth of theorizing and research about communication as dialogue (see, for example, Anderson, Cissna, & Arnett, 1994; Buber, 1958; Habermas, 1979, 1984, 1987; Johannesen, 1971; Pearce & Littlejohn, 1997). In general, dialogue is defined by an open and honest exchange of ideas between or among actors who agree to suspend judgment, speak honestly, and remain profoundly open to change (Eisenberg, Goodall, & Trethewey, 2006). Despite the utopian nature of this definition, experience has taught us that dialogues do not occur in a vacuum. Actors bring to even the most open and honest dialogic events a context made of their own historical, cultural, religious, and political sense-making *schema*. In other words, actors interpret the meanings of their message exchanges through an emerging, flexible framework constituted in language through an ongoing process of retrospective sense-making (Pearce & Pearce, 2000a; Weick, 1995), current analyses of meanings (Spano, 2001; Kellett & Dalton, 1999), and future projections of goals, plans, self-interest, ego, and needs (Pearce & Littlejohn, 1997; Pearce & Pearce, 2000b; Eisenberg, 1984). Because of the complex and uncertain nature of dialogic encounters, adopting the idea of dialogue for strategic communication purposes in the Middle East is probably unwise.

Even if it were possible, in the spirit of the Camp David Accords,[1] to bring together U.S. and Muslim political leaders for the purpose of dialogue, the prospect has been seriously compromised by recent public diplomacy and credibility failures (see Goodall, et al., 2006b). By seeking political influence over cultural and religious understanding, by relying on a one-way model to inform communication campaigns, and by our perceived misuses of military might to accomplish strategic objectives without exhausting diplomatic possibilities, we may have forfeited any real opportunity for dialogue with the Muslim world at the highest levels of government.[2]

According to Jurgen Habermas in his theory of communicative action (1979), dialogues are *symmetrical interactions* characterized by "reciprocal expectations regarding the truth, appropriateness, and sincerity of statements" as well as an "openness to being persuaded through the process of communication" (Dutta-Bergman, 2006, p. 104). *Where these conditions for symmetry cannot be met, there can be no dialogue*. Until the U.S. rebuilds its credibility, reestablishes *trust* with other nations and leaders, and recommits itself to *truth* as well as effectiveness as our standard metrics for strategic communication efforts, those interested in improving our image must look to less grand ideas *in the middle ground of communicative practices between monologue and dialogue* that have a better likelihood of success.

Inability to Prepare For, or Respond To, the Global Jihadi Media Strategy

The second major reason for U.S. public diplomacy failures in the Middle East is the rise of Internet influence on Muslim (and particularly jihadi) public opinion (Brachman, in press; Combating Terrorism Center, 2006; Hunt, 2006; International Crisis Group, 2006; Hoffman, 2004). Studies have documented the U.S. government's inability to prepare for, or respond to, the sophisticated jihadi media strategies that have successfully captured the imaginations of many people—and particularly the youth—in the Middle East (see Arquilla & Ronfeldt, 2001; Corman & Schiefelbein, 2006; Hunt, 2006; Nisbet, et al., 2004; Rumsfeld, 2006).

Corman & Schiefelbein (2006) provide an analysis of the three core communication strategies embodied in jihadi websites and media: the *legitimation* of the global jihadi movement within existing social and religious frameworks commonly understood in the region; the *propagation* of that message to sympathetic audiences in regions where the movement seeks recruits and political expansion opportunities; and the use of *intimidation* to cow opponents as well as those within the Muslim world who may turn against them.

The new media made possible by access to the Internet and traditional media outlets require a more pragmatic, middle-ground effort known as "strategic ambiguity."

What is needed to counter the coordinated media campaign of jihadi groups is a coordinated strategic communication plan organized, understood, and deployed consistently by the U.S. and its allies. Moreover, this plan should combine "a long-term strategy for improving our credibility with Muslim audiences" (Corman & Schiefelbein, 2006, p. 2) with an active engagement of issues pertinent to local audiences both within the Muslim community at home and abroad. Until the U.S. sufficiently organizes its own Muslim resources to combat the advances made by jihadi groups, there is little likelihood of global success.

Here again, the communication principles driving these counter-blogging, counter-media campaigns cannot be derived from an outdated, ineffective monologic approach, nor can it credibly rely on dialogic models. The new media made possible by access to the Internet and traditional media outlets require a more pragmatic, middle-ground effort known as "strategic ambiguity" (Eisenberg, 1984; Eisenberg, Goodall, & Trethewey, 2006).

Strategic Ambiguity as a Model for Strategic Communication

Strategic ambiguity is a mid-range theory that is easily adapted to the purposes of improving the communication process necessary for rebuilding the U.S. image abroad and furthering our diplomatic objectives on a global mediated stage. Strategic ambiguity theory is drawn from research about building resilient organizations in turbulent environments under conditions of uncertainty.[3]

"Message control" as an organizing metaphor should be replaced by something that allows for rapid dissemination of information and flexibility in response to the diverse needs of a global workforce and marketplace.

The idea of strategic ambiguity as a communication strategy emerged in the mid-1980s. At that time, flattened business hierarchies made possible by new information technologies coupled with the need to be faster and more responsive to global markets challenged existing top-down models of information sharing and communication in organizations (Eisenberg, 1984). The old organizing model—much like the monologic model for communication—was informed by a "control" metaphor that itself was firmly rooted in the assumption of a shared organizational culture that respected hierarchies of power, strict divisions of labor, and the power of the higher authorities in the company to direct work activities as well as their meanings.

By contrast, organizational theorists and enlightened business leaders posited that "message control" as an organizing metaphor should be replaced by something that allows for *rapid dissemination of information* and *flexibility* in response to the diverse needs of a global workforce and marketplace. An alternative organizing schema and message strategy rooted in "strategic ambiguity" rather than message control enabled a much wider sharing of information necessary for employees and customers to make better decisions as well as allowed for "local empowerment" of meanings associated with the implementation of vision, mission, values, and goals (Eisenberg, Goodall, & Trethewey, 2006).

Strategic ambiguity as a communication strategy occupies a theoretical middle ground between monologic control and dialogic empowerment models. Strategic ambiguity values the symbolic and dialogic nature of language and the multicultural bases for interpretations of meanings. It also values what Eric Eisenberg terms "unified diversity" so necessary to the creation of resilient organizations operating in highly uncertain environments (Weick & Sutcliffe, 2001). Yet this model does not operate without guiding principles capable of informing a coordinated management of meaning across diverse audiences (Pearce and Pearce, 2000; Pearce, 1989). In the final section of this chapter, we articulate those principles.

Five Pragmatic Principles to Guide Policy

Strategic ambiguity operates between monologue and dialogue in the public sphere. Given the failures of the monologic model and the unlikelihood of dialogue between or among disparate leaders operating at the highest levels of diverse governments, strategic ambiguity is a viable and appropriate model for engaging public diplomacy in an uncertain world.

The goal of strategically ambiguous communication should not be "shared meaning" but instead "organized action." Five principles to guide a new communication policy are:

1. **Practice strategic engagement not global salesmanship:** *Strategic engagement is the application of strategic ambiguity to public diplomacy goals.* Demonstrate a willingness to engage the messages of other leaders and spokespersons without seeking immediate closure or insisting on the inherent "rightness" of our messages. Consider communication a two-way interaction and meanings to be emergent over time.[4]
2. **Do not repeat the same message in the same channels with the same spokespersons and expect new or different results:** *Repetition breeds contempt.* Using a monologic model to inform public diplomacy doesn't work because it encourages spokespersons to repeat the same basic ideas

The goal of strategically ambiguous communication should not be "shared meaning" but instead "organized action."

and messages without accounting for the meaning-making practices of intended audiences. Replace repetition with strategic engagement as a guiding principle of communication with diverse audiences.

3. **Do not seek to control a message's meaning in cultures we do not fully understand:** *Control over preferred interpretations is a false goal in a diverse mediated communication environment.* There is an inverse relationship between control over a message's meanings and our understanding of the cultures wherein it will be interpreted. The less we know about cultures, languages, and religions, the less control we can fairly exert on the likely meanings attributed to public diplomacy messages.
4. **Understand that message clarity and perception of meaning is a function of relationships, not strictly a function of word usage**: *Focus diplomatic efforts on building trust and credibility based on a longer term and deeper understanding of cultures, languages, and religions.* It is in the context of building ongoing relationships and being responsive to the interpretations of others that we are able to craft productive messages that have legitimate value and that resonate among diverse audiences.
5. **Seek "unified diversity" based on global cooperation instead of "focused wrongness" based on sheer dominance and power:** *Recognize that shared meaning isn't the only goal, but shared principles and goals are singularly meaningful.* Learn to expect and cultivate multiple meanings in local cultures and communities that support broader agreements of U.S. principles or goals and expect that those audiences will adapt and internalize those messages according to their own needs and resources. Building coalitions of engaged communication should actively and publicly augment coalitions of military force and be coordinated with them.

Notes

1. See the framework for this historic dialogue at: *http://www.jimmycarterlibrary.org/documents/campdavid/accords.phtml.*
2. This fact should in no way diminish current and future dialogue projects sponsored by the Department of State, USAID, or private consortiums dedicated to improving understanding and facilitating productive change through dialogue. One excellent model for how dialogue can be used to bring together diverse groups is the Consortium for Public Dialogue, which sponsors a wide variety of these events

annually (see their website: *http://www.publicdialogue.org/projects/index.html*).

3. In some diplomatic conversations, the term "strategic ambiguity" has been used as a derogatory term to refer to the administration's unwillingness to stake a fixed and unambiguous policy regarding the independence of Taiwan from China. Our use of the term is not in any way related to that limited characterization or issue.
4. See excellent overview in Banks, Ge, & Baker (1991). For example, the U.S. failed a major political opportunity in U. S.–Egyptian relations when LBJ was offended at what he thought were bellicose comments by President Nasser. Nasser later explained his comments were only aimed at audiences at home while he wanted to work in diplomatic backchannels with the U. S. Understanding sedimented cultural norms and values and their impact on communication patterns requires an engaged posture, not a "our way or no way" posture. A more recent example may be found in Goodall, et al., (2006b) in the U.S. mishandling of diplomacy involved in responding officially to the letter written by Iranian President Ahmadinejad to President George W. Bush.

References

Anderson, R. Cissna, K. N. & Arnett, R. C. (1994). *The reach of dialogue: Confirmation, voice and community.* Cresskill, NJ: Hampton Press.

Arquilla, J., & Ronfeldt, D. (2001). The Advent of Netwar (Revisited). In J. Arquilla & D. Ronfeldt (Eds.), *Networks and netwars: The future of terror, crime, and militancy* (pp. 1–25). Santa Monica, CA: Rand.

Banks, S. P., Ge, G., & Baker, J. (1991). Intercultural encounters and miscommunication, In N. Coupland H. Giles, & J. M. Wiemann, (Eds.). *"Miscommunication" and problem talk* (pp. 103–120). London: Sage Publications.

Berlo, D. (1960) *The process of communication*. New York: Holt, Rinehart and Winston, Inc.

Brachman, J. (in press). Al-Qaeda: Launching a global islamic revolution. *The Fletcher Forum of World Affairs*.

Buber, M. (1958). *I and thou.* New York: Macmillan.

Combating Terrorism Center (2006). *Harmony and disharmony: Exploiting al-Qa'ida's organizational vulnerabilities*. Report by the Combating Terrorism Center, United States Military Academy. Feb. 2006. Available online: http://www.ctc.usma.edu/aq.asp

Corman, S. R., & Schiefelbein, J. S. (2006). *CSC Report #0601—Communication and Media Strategy in the Jihadi War of Ideas*. Report from the Consortium on Strategic Communication, Arizona State University, April, 2006.

Corman, S. R., Hess, A., & Justus, Z.S. (2006). *CSC Report #0603—Credibility in the Global War on Terrorism: Strategic Principles and Research Agenda.* Report from the Consortium on Strategic Communication, Arizona State University, June 9, 2006.

Dutta-Bergman, M. J. (2006). U.S. Public diplomacy in the middle east: a critical cultural approach. *Journal of Communication Inquiry, 30,* 102–124.

Eisenberg, E.M. (1984). Ambiguity as strategy in organizational communication. *Communication Monographs*, 51, 227–242.

Eisenberg, E. M., Goodall, Jr., H. L., & Trethewey, A. (2006). *Organizational Communication: Balancing Creativity and Constraint, 5th ed.* New York: St. Martin's Press.

Goodall, Jr., H. L. (2006). Why we must win the war on terror: communication, narrative, and the future of national security. *Qualitative Inquiry, 12*, 30–59.

Goodall, B., Cady, L., Corman, S.R., McDonald, K., Woodward, M., & Forbes, C. (2006). *CSC Report #0602—The Iranian Letter to President Bush: Analysis and Recommendations.* Report from the Consortium on Strategic Communication, Arizona State University, May 18, 2006.

Goodall, Jr., H. L., and Goodall, S. (2006). *Communication in Professional Contexts: Skills, Ethics, and Technologies,* 2nd ed. Belmont, CA: Thompson Learning/Wadsworth, 2006.

Habermas, J. (1979). *Communication and the evolution of society.* Boston: Beacon.

Habermas, J. (1984). *The theory of communicative action, Vol. 1: Reason and the rationalization of society*. Boston: Beacon.

Habermas, J. (1987). *The theory of communicative action, Vol. 2: A critique of functionalist reason.* Boston: Beacon.

Hayakawa, S. I. (1978). *Language in thought and action.* (4th ed.). New York: Harcourt, Brace, Jovanich.

Hoffman, B. (2004). The changing face of Al Qaeda and the global war on terrorism. *Studies in Conflict & Terrorism, 27*, 549–560.

Hunt, K. (2006, March 8). Osama bin Laden fan clubs build online communities. *USA Today.* Available online: http://www.usatoday.com/printedition/news/20060309/a_google09.art.htm

International Crisis Group (2006, February 15). *In Their Own Words: Reading the Iraqi Insurgency*. Middle East Report N°50. Available online: http://www.crisisgroup.org/home/index.cfm?id=3953&l=1

Johannesen, R. L. (1971). The emerging concept of communication as dialogue. *Quarterly Journal of Speech, 57,* 373–382.

Kellett, P. M., Dalton, D. G. (1999). *Managing conflict in a negotiated world.* Thousand Oaks, CA: Sage.

Littlejohn, S. W., & Domenici, K. (2000). *Engaging Communication in Conflict: Systemic Practice.* Thousand Oaks: Sage.

Nisbet, E. C., Nisbet, M. C., Scheufele, D. A., & Shanahan, J. E. Al. (2004). Public diplomacy, television news, and Muslim opinion. *Press/Politics, 9*(2), 11–37.

Pearce, W. P. (1989). *Communication and the human condition.* Carbondale: Southern Illinois University Press.

Pearce, W. P., & Littlejohn, S. (1997). *Moral conflict: when social worlds collide.* Thousand Oaks: Sage.

Pearce, K. and Pearce, W. P. (2001). The public dialogue consortium's school-wide dialogue process: A communication approach to develop citizenship skills and enhance school climate. *Communication Theory, 11*, 105–123.

Pearce, W. P., & Pearce, K. (2000a). Combining passions and abilities: Toward dialogic virtuosity. *Southern Communication Journal. 65,* 161–175.

Pearce, W. P., & Pearce, K. (2000b). Extending the theory of the Coordinated Management of Meaning (CMM) through a community dialogue process. *Communication Theory, 10,* 405–423.

Rumsfeld, D. (2006). *New realities in the media age: a conversation with donald rumsfeld.* Council on Foreign Relations. Available online: http://www.cfr.org/publication/9900/

Spano, S. (2001). *Public Dialogue and participatory democracy: The Cupertino Project.* Cresskill, N. J.: Hampton Press.

Weick, K. (1995). *Sensemaking in organizations.* Thousand Oaks, CA: Sage.

Weick, K., and Sutcliffe, K. (2001). *Managing the unexpected: Assuring High performance in an age of complexity*. San Francisco: Jossey-Bass.

CHAPTER THREE

Leadership Reconsidered as Historical Subject

Sketches from the Cold War to Post-9/11

ANGELA TRETHEWEY AND H. L. GOODALL, JR.

Introduction

In this essay, we seek to further an understanding of leadership as the evolution of perspectives embedded within a wider historical frame. It is our contention that recovering the historical, economic, and cultural basis for leadership theories since WWII reveals three hidden storylines; each one of them tied to the three dominant historical eras. By hidden storylines we refer to the subtexts and the forgotten and lesser known, but no less foundational aspects of the more popularized and codified leadership theories that came to define each of three historical eras. Those three eras are the *Cold War*, the *Post-Cold War* and the *Post-9/11* era.

Understanding the emergence of the "dark side" of leadership in a post-9/11 environment is intricately tied to uncovering and narrating the relationship of ideas about leadership to the dominant material and ideational struggles that defined each of these three historical eras. Recovering the "bright side" of leadership as an oppositional narrative and reversing the trend toward fundamentalism require a new model of leadership. We offer "pragmatic complexity" as a provisional step toward that new model.

Background: The Need to Reframe Leadership

Theories of leadership provide a story that is largely ahistorical. Divorced from the social and cultural discourses that shaped them, disconnected from the political and economic realities that surrounded their making, and seemingly immaculate in their conception as ideas, these free-floating signifiers we call theories of leadership are the bastard children of all that has been omitted from their lineage.

We are interested in a larger project that examines organizing and identity as products of a post-9/11 discourse. In this chapter, we limit our discussion to leadership. We take an alternative view of leadership situating the theories we associate with it within a grand narrative of post-world war II history and culture. Both communication and management scholars are calling for analyses that trace current and often taken-for-granted understandings of foundational concepts, including gender and organization, from their historical roots. Such holistic and historic analyses have the "potential to hone our understanding of the linkages among discourse, history, and material conditions" (Ashcraft & Mumby, 2004, p. 23). As Collinson and Grint (2005) sum it up:

> Leadership 'research' has frequently been at best fragmented and at worst trivial, too often informed by the rather superficial ideas of management and academic consultants keen to peddle the latest prepacked list of essential qualities deemed necessary for individual leaders and as the proscribed solution to all leadership dilemmas. Within business schools and management departments leadership has often remained a "Cinderella" subject, neglected and/or underestimated by those keen to analyze and theorize the social, political, organizational, and philosophical dimensions of human affairs. Consequently, the intellectual integrity of leadership as a legitimate and important field of study has remained open to question. (p. 5)

Indeed, the rapid proliferation and subsequent replacement of proven and essential leadership strategies in the popular management press suggests that simple solutions to complex leadership problems rarely work (Pfeffer & Sutton, 2006). We argue that it is precisely because leadership scholars and practitioners too often rely on the aforementioned essential individual qualities that they fail to understand how leadership is part and parcel of a larger, complex socially constructed and uniquely historicized discursive reality.

Among management and organizational theorists there are those who have taken seriously the historical milieu that shaped current understandings of, for example, managerial ideologies of control (Barley & Kunda, 1992), the discursive construction of "the employee" (Jacques, 1999) or the discursive construction of the "airline pilot" (Ashcraft & Mumby, 2004). However, these analyses are the exception rather than the rule. As James McGregor Burns poignantly argues (2005), "We are all indebted to historians and other scholars who minutely examine specific events and problems in the hope that the devil is indeed in the details. But these scholars often fail to place the details in a broader chain of causality or even a frame of reference" (p. 12). Scholars of organizing are poised to conduct historical analyses in order to locate central ideas, like leadership, in this broader context.

The present study seeks to further an understanding of leadership as the evolution of perspectives embedded within a wider historical frame. It is our contention that recovering the historical, economic, and cultural basis for leadership theories since WWII reveals three hidden storylines; each one of them tied to the three dominant historical eras. In this chapter, we trace how these hidden storylines became the often unspoken absences that provided discursive resources for "darkside" leaders to emerge as a powerful presence in ways that were not entirely inconsistent with their "brightside" counterparts. By hidden storylines we refer to the subtexts and the forgotten and lesser known, but no less foundational aspects of the more popularized, commodified and dominant leadership theories that came to define each of three historical eras. Those three eras are the *Cold War*, the *Post-Cold War* and the *Post-9/11* era. We believe that studying each of these eras offers a unique perspective on the kinds of ideological foundations critical to the development of theories of leading and organizing. In addition to the inherent value of exploring those relationships in time is the need to more fully appreciate the reasons why earlier models of leadership neither accurately account for actual performance nor adequately inform current practices (Collinson, 2005). Understanding the emergence of leadership in a post-9/11 environment is intricately tied to understanding the relationship of ideas about leadership to the dominant material and ideational struggles that defined each of these historical eras.

One *caveat* informs our intentionally general observations about the evolution of leaders since WWII. First, it is both easy and often unwise to be seduced by the neatness and tidiness of constructed analytic categories. We know that the history of leadership—just as history itself—is rife with ruptures, discontinuities, multiple interpretations, and competing narratives engaged in hegemonic struggles.[1] Nevertheless, the theoretical benefit of constructing general categories enables us to write a history of the present (Foucault, 1979) that allows us to uncover the historical traces and dominant trends informing the language of leadership across the discourses of various organized communities (e.g., politics, business, education, etc.).[2]

We believe that recovering these histories benefits our understanding of the evolution of leadership in three ways. First, it broadens and deepens our appreciation of the relationship between grand cultural narratives and everyday organizing practice. Second, it affords a new view of the evolution of leadership in relation not just to academic and practical concerns but in relation to the ideological and dialectical struggles that mark those times (Collinson, 2005). Finally, our analysis demonstrates that traces and perversions of earlier "bright side" ideas of leading and organizing are characteristically redefined, appropriated and perverted in darker times of crisis. In so doing, this chapter contributes to a small, but growing body of research on the "dark side" of leadership (Blase & Blase, 2002; Johnson, 2001; Lutgen-Sandvik, 2003)

All leadership is political (Mumby, 1988). Political leadership has always been defined in relation to crisis, whether it was an ideological struggle between the world's great superpowers, or the crisis born of the victory of capitalism and the demise of the former Soviet Union and its satellites, or the struggle born of misreading religious, economic and cultural differences that have led us into the Global War on Terror (GWOT). The reluctance to read the political implications of global crisis as background for the emergence of dominant models of leadership in the business world and leadership literature has contributed not only to a lack of historical knowledge of leadership as historical subject but also to an unwillingness to acknowledge the dangerous fallout from those troubled times.

Organizing the Leadership Narrative: Textual Fragments and Discursive Themes as Method

There are many ways to tell this story; indeed, there are many stories that have yet to be told about leadership in a post-World War II history. Each one of them is undoubtedly partial, partisan and problematic (Eisenberg, Goodall, & Trethewey, 2007). We make no claim that the story we're about to tell is *the* history, it is merely *a* history shaped by our experience, reading and work as communication scholars of organizational life. The choice we have made is to simply provide a narrative framework that delimits the three historical periods that shape our contemporary understanding and theorizing of leadership.

The method we use to delineate the eras and their leadership themes borrows from postmodern rhetorical strategies. In particular, following McGee (1990), we are interested in "building a text worthy of criticism" out of cultural fragments. In this case, our fragments are popular books about leadership and the historical/cultural/political surround within which they emerged. We believe that these fragments trace the contours of popular understandings of leadership, specifically "how particular practices and concepts become accepted at certain historical periods as being natural, self-evident, and indispensable" (Townley, 1994, p. 2; see also Knights & Willmott, 1999).

How did we do this? We chose central texts for each era that provided us with emblematic representations of dominant cultural beliefs about the (seemingly innate and taken-for-granted, but always discursively constituted) "nature" of leadership and leading as well as a larger cultural discourses about organizations, employees, the meaning of work, and the construction of an ideal business self, or in this case, an archetypal leader (Czarniawska-Joerges & Wolff, 1991). By "central" we mean both best-sellers and influential texts for practicing managers and scholars of management and organization.

For the Cold War Era, we focus on the texts of Abraham Maslow, whose "hierarchy of needs is ubiquitous in management education and theory" (Cullen, 1997, p. 355) and whose work serves as an often taken-for-granted foundation for contemporary management and leadership theory and practice (Dye, Mills & Weatherbee, 2005; Grant & Mills, 2006; Runté & Mills, 2006). His story could have been told differently. Maslow was a complex, layered, and sometimes counter-intuitive theorist who work has been reduced to simplistic prescriptions. Other textual fragments are drawn from the work of Cold War-era theorists, including McGregor (1960), Whyte (1956), Blake and Mouton (1964), Hersey and Blanchard (1969), and others.

The theorist who serves as our exemplar of the post-Cold War era is Jim Collins. His text, *Built to Last: Successful Habits of Visionary Companies*, co-authored with Jerry Porras, is the most influential management book since the fall of the Berlin Wall (Ackman, 2002). And his book, *Good to Great: Why Some Companies Make the Leap . . . and Others Don't* rounds out the list of Forbes's 20 Most Influential Business Books from 1981–2000 (Ackman, 2002). Given the proliferation of management and leadership popular press books and other "products" in this era, several other influential texts serve as textual fragments that help us to construct our narrative (e.g., Bennis, 1994; Covey, 1992; Porter,

1998a, 1998b; Welch, 2001, 2005).

The historical narrative of the post-9/11 era is, of course, emerging as we write and is not yet as easily contained in an "iconic" or exemplary text. Our reading of recent leadership texts indicates that in a post-9/11 environment, leadership theories are centered on how to deal with an increasingly ambiguous and uncertain global environment. We trace two narratives, one dominant and one oppositional, that have emerged in response. First, from the right, there is a dominant fundamentalist rhetoric rooted in an understanding of leadership as quest for ultimate control and security (Taylor, 2005; Parrett, 2007). From the left, there is an oppositional view of how to respond to uncertainty using ambiguity as strategy (Eisenberg, 2007), counter-intuitive resources for creating enlightened understandings of systemic causes and effects (Gladwell, 2002, 2007), and a willingness to decouple leadership from control and embrace "pragmatic complexity" (Corman, Trethewey, & Goodall, 2007).

At present, there is no clear victor in this discursive struggle, but as we have seen in the Cold War and post-Cold War eras, the narrative that is easiest to re-tell and popularize is likely to be equated with the iconic weight of the "truth" of leadership theory and practice for our era. Our task must be to remain vigilant and not succumb to the temptation of "simple is best" when it comes to explaining leadership. We must become actively involved in writing and re-telling the story of leadership as both an historical narrative and a rhetorical vision for the future. To begin this project, we now turn to the recovery of a hidden storyline that traces the "darkside" of leadership from the Cold War to our post-9/11 era.

The Cold War Era Leadership Theories: The Era of Self Actualization

One great advantage of writing about leadership within the Cold War era is that we know how the Cold War turned out, or at least how the dominant cultural *story* ended (see Gaddis, 2005; Goodall, 2006a; Judt, 2005). We also have a clear and virtually consensual record of the dominant ideas about leadership that defined the 1945–1991 era. Indeed, our own discipline of communication was born in World War II as "basic communication(s) skills" were needed to transform millions of heretofore-untrained civilians into industrial workers or military specialists (Redding, 1985, p. 25). It was during the Cold War years, however, that communication, as a discipline, found its footing in service of the new and burgeoning military-industrial-academic complex (Redding, 1985). Any quick examination of organizational or management texts providing a history of leadership during that era tells a story that begins with traits and styles of leading that were informed by human relations and (later) human resources approaches to understanding human motivation. These early descriptors of leaders and leadership were then augmented by emergent notions of contingencies and situations that eventually led to the development of ecological systems theories and the rise of performative cultures (Barley & Kunda, 1992). It is important to note, however, that new ideologies do not replace old ones entirely; rather, new leadership narratives are overlaid onto older ones, creating ideologies that are not entirely seamless, contain ruptures, contradictions, and create the space for struggles over meanings of leadership across historical eras (Holmberg & Strannegard, 2005).

Central to all of these advances in leadership theorizing was the essential idea of the human as a questing agent, "rotten with perfection" as Kenneth Burke had it, and driven (when not overtly by economic and political interests) by the sometimes contradictory goals for the questing self of conformity and spirituality. Chief among the theorists most closely associated with this utopian thought was Abraham Maslow (Grant & Mills, 2006). We call this "utopian thought" because Maslow believed that human organization and the leadership self were *perfectable* and that good management led to healthy individuals who would use their creative capacities to further not only the goals of the organization but of themselves. The goal of every worker was to become liberated through developmental stages of awareness and action, which were enabled by visionary leaders with an appreciation of humans as questing souls. In his view, on the "path to success" in *Eupsychian Management*:

> . . . it is well to treat working people as if they were high type Theory Y human beings, not only because of the Declaration of Independence and not only because of the Golden Rule and not only because of the Bible or some religious precepts or anything like that, but also because this is the path to success of any kind whatsoever, including financial success. (1963)

As Bill Cooke, Albert Mills, and Elizabeth Kelley (2005) argue, Maslow's notion of enlightened leadership was in direct response to the Cold War in both material and ideational ways. Ideologically, Maslow—like his academic counterparts interested in human relations and human resources—was driven by an elite utopian vision that would triumph when humankind was freed from the forces of totalitarianism, whether in the form of Communism or Fascism. His hierarchy of needs posits an ever-upward progress narrative shaped by the desire to self-actualize that itself is rooted firmly in the accomplishment of safety, security, belonging, and the development of healthy self-esteem.

In addition to Maslow, this era produced advances in leadership theorizing and research that reflected the utopian desire for continuous improvement based on the application of thoughtful science to the practical problems of leading. Douglas MacGregor's (1960) Theory X and Theory Y highlighted the natural superiority of an enlighted, empowering human relations oriented leader; Fred Fiedler (1967) studied what effective leaders actually did and discovered that anticipating and responding to contingencies in their environments determined their success; John Hersey and Paul Blanchard (1977) found that there was a pattern of directing and supporting actions that led teams to successful outcomes, and so forth. Moreover, the theme that workers should continually and purposefully strive for success worked its way into mass-market "how-to-succeed" texts written for middle-class men, who were not necessarily leadership material. Those men were encouraged to "get along" with their superiors. "This formula suited the needs of corporate [and cultural] life in the 1950s and justified the lives of 'organization men'" (Biggart, 1983, p. 302).

The overarching utopian backstory for these advances underscored two essential truths: (1) the purpose of leadership was both the accomplishment of the organization's goals and self-actualization on the part of the leader and followers; and (2) the method for accomplishing those goals was rooted in a developmental pattern that pitted "brightside leaders" (Theory Y, Self-Actualizing, Contingent, Team-based Empowerers) against their "darkside" counterparts (Theory X, rigid, control-oriented, repressive regime masters). If there was ever a clearer symbolic depiction of Capitalism versus Communism, of the US and our allies against the Soviets and their satellites, we cannot name it.

So, given these Cold War "the world is at stake" terms, what kind of business leader was called for under these circumstances? What kind of organizational communication was believed to be necessary? In times of great conflict (between ideologies and superpowers), what do people rhetorically crave? Judging from experience in North America, we can offer the following observations:

- A reverent, yet secular leader with heroic qualities who has a self-actualized, democratic style of leading and is "better" than those he (or, rarely, she) is charged with leading (White & Lippett, 1960);
- A leader with the empirical traits of an authoritative plain-speaking citizen who always links Capitalism to patriotism and who inspires others to understand the need to do one's duty through hard work

personal sacrifice, and teamwork in service to a cause greater than oneself (White & Lippett, 1960);

- A top-down communicator who expects underlings to conform, fit in, and to emulate superiors in an *Organization Man* fashion and who balances a need to complete tasks with a need to satisfy relationships (Whyte, 1956; Blake & Mouton, 1964, 1985);
- A "doer" (e.g., telling, selling, participating, delegating) rather than a "thinker" (Hersey & Blanchard, 1969).

As time and ideas about organization and leadership traits evolved during this Cold War period through systems and into cultural approaches, the characteristics we have identified above also morphed into newer rhetorical forms. For example, strong cultures were closely associated with charismatic leaders whose "betterness" and elite status was often a by-product of their authoritative yet symbolically brazen style. Similarly, the evolution of a plain-speaking ordinary citizen who attained a leadership position through good work was culturally supplanted by more of a performative and consumptive model. Advancement was attained through personal networks and proven salesmanship. The iconic leader was a zealot for the enterprise, a la Deal and Kennedy (1982) and Peters and Waterman (1983).

Yet there were, among these advances in bright side leadership theories, traces of perversion: Capitalist cultures and leaders began to link peak experiences and self-actualization to the material world of markets and commodities. By the time of the collapse of the Soviet empire, we had succeeded in replacing a spiritual nirvana with the affluent accumulation of stuff.

Post-Cold War Leadership Theories: The Era of Spin

Following the breakup of the former Soviet Union, Capitalism signaled its world-conquering victory with an unprecedented expansion of the free market economy (Jansen, 2005). Globalization became the dominant force in shaping re-organization and re-thinking business practices. In our brave new hyper-capitalist world, citizenship (an organization of the idea of self around principles of shared responsibility for democracy) was replaced by rampant consumerism (an organization of the self around principles of individual ownership) and market rationalization (Kunda & Souday, 2004). Technology, the cool-medium handmaiden to Capitalism, accelerated the replacement of goods with services, production with reproduction (du Gay, 1997).

One result was the creation of a cultural need for an iconic, media-savvy leader capable of uniting and inspiring audiences world-wide through scripted, focus room, and poll-tested messages in an unending campaign for product placement and popularity known as "branding" (Lair, Sullivan & Cheney, 2005). Virtual worlds and their glittering surfaces beckoned those who believed in their own entitlement as much as the previous generation believed in sacrifice: Stock markets bubbled, the world economic party was never over for the "haves" just as hard work at low pay was never over for the "have nots"; and for those of us in the privileged classes in the affluent northern hemisphere, anything seemed possible as we rapidly approached the birth of a grand new millennium.

It is who we did not see, what we were not paying attention to, and who we had always marginalized, colonized, exploited, and ignored who were then organizing against us, against all that the West and Godless Capitalism stood for, and all that was symbolized by Washington, D.C., the Pentagon, and the Twin Towers of the World Trade Center.

But we're getting ahead of our story.

What kind of business leader was called for under these prosperous and seemingly limitless circumstances? What kind of organizational communication was necessary? In times of great prosperity and potential, what do people rhetorically crave? Judging from experience in North America, we can offer the following observations:

- A leader is an inspirational "super-salesman," a wo/man for all markets, a made-for-TV-consumption image armed with a vision, a mission, goals, and a strategic plan (Bennis, 1992; Peters, 1999);
- Communication is an advertisement for the self; symbols of identity replace character as a fluid, market-driven performance dispersed across multiple audiences, theatres, and media (Bolman & Deal, 1991);
- Cult-like cultures (Collins & Porras, 1994) replace traditional authoritarian notions of the organization where charisma trumps substance;
- Progressive leaders were advised to think of, and to position themselves as "stewards" of their enterprise and the environment; forward-thinking leaders were tasked with creating "learning organizations" (Senge, 1991) and employees across the board in progressive organizations were taught to think of themselves as democratic stakeholders (Deetz, 1995; Wheatley, 1992) and collaborators (Holmberg & Strannegard, 2005).

On the bright side of leadership theorizing during this post-Cold War era, a virtual parade of management gurus touted the value of *transformative* knowledge and action. Central to the transformative leadership style was a passion to make a positive contribution to the world through organizational empowerment and team-based learning. Personal capacities for change and growth, deep spiritual values, and ethical commitments to the fair treatment of others dominated the rhetorical and narrative surfaces of these advisories.

Yet there was also a dark side to the transformative approach to leading. For one thing, while change and ambiguity may be constants in our world, predicting the future based on a whiz-bang consulting model is often tricky business. Pfeffer and Sutton (2006) argue that this model of leadership has led to a market-based, entrepreneurial ideology that substitutes individual gut-level, symbolically rich decision making for decisions based on empirical observations or substantive contributions. Not surprisingly, post-Cold War leaders moved from one management fad to another, rarely attending to the consequences of their actions (Zorn, Page, & Cheney, 2000). So, we witnessed leaders moving through the "seven habits" (Covey, 1992) to "awakening the giant within" (Robbins, 1992) to "finding the cheese" (Johnson & Blanchard, 1998). Like the gurus who forward these ideas, leaders of this era were encouraged to "spin" their ideas, their decisions and their identities according to the latest fashion. Furthermore, in a consumer-rich culture where charisma trumps character for awe-inspired managers and citizens and spin replaces truth for shareholders and voters, short-term economic and political gains rather than long-term ecological investments and decision making begin to make perfect, if totally corrupt, global sense. Moreover, the systematic exploitation of workers through outsourcing in Third World countries allows companies to remain competitive while downsizing home labor forces, thereby transferring to the government the responsibility for health care, education, and pension funds that once were staples of the labor contract (Porter, 1998a, 1998b). What may be good for General Motors becomes bad for the country.

Post-9/11 Leadership Theories: The Emergence of Fundamentalism

Following the tragic attacks on the World Trade Center and the Pentagon, leaders worldwide found themselves confronting not only a new non-state enemy (terror) but moreover a challenging organizational problem: how to protect and defend against fear, anxiety, uncertainty, and the likelihood of a terrorist incident. Primary among business decision-making was the idea of disaster preparedness, knowledge management, and disaster recovery. In strict business terms, calculating the real number of minutes (or, God forbid, hours) that markets would be unavailable or inventories disrupted in the event of a nuclear, chemical, or biological attack has taken precedence over plans for preventing death or injury to workers. Additionally, insuring that the core of the business enterprise remains intact depends on creating, tracking, cataloguing, distributing, and selling knowledge (Awad & Ghaziri, 2003; Iverson & McPhee, 2003). Thus, surveillance becomes a primary organizing process and tracking, cataloging, managing and displaying knowledge (or intelligence) products are increasingly central activities of both corporations and the government, particularly in a post-9/11 environment (9/11 Commission Report, 2005).

Interestingly, corporations now routinely make use of some of the same intelligence-gathering tactics that government agencies do and the lines between them are becoming fuzzier. To wit, Target Stores now has the largest "Assets Protection" program in the U. S., the third largest and perhaps the most state-of-the-art forensics lab in the country (second only to the FBI's), and routinely "partners" with local law enforcement (e.g., the newest Target store in our area houses a local police precinct and is reinforced by bullet-proof storefront windows). Target Stores employs a host of knowledge management technologies to track its customers and other potential "risks." For example, if you were to purchase more than the recommended dosage of Sudafed at Target, cameras would be instantly trained on your face and your image would be beamed to local police departments to determine whether or not you are a suspected methamphetamine dealer. Target also asks its potential employees to "See Yourself" in a career in "investigations" where new recruits "plan, develop and lead investigations to address financial risk at a market or national level. With available data and intelligence, you will identify significant risks and implement investigations and tactics to address them" (*http://target.com/target-corp_group/careers/assets_protection.jhtml*). Left unchecked, this need to manage information can lead to cultures grounded in increased secrecy, surveillance, and panoptic control (Trethewey & Corman, 2001).

This post-9/11 environment has created discursive conditions that enable a particular sort of organizing that has come to characterize the state, corporate and educational institutions. Those institutions are responding to a rise in ambiguity that is now experienced as anxiety and uncertainty about the future that is reinforced and reproduced or "cultivated" constantly in a saturated 24/7 media environment (Gerbner et al., 1986; Goodall, 2006b). Insecurity produces a need for security often at the expense of civil and employee rights.[3]

In the post 9/11 business environment leadership has also changed. The political has become capital.[4] In an inverse of Maslow's projected path toward business and personal success, leaders seem to have retreated back down the hierarchy of needs to protecting our security, insuring our safety, and worrying about our ability to provide food and shelter to loved ones. Indeed, since 9/11, popular models of leadership rely much less on an "inspirational" model of leadership and more on a "connectionist" model. This new leader is not necessarily heroic, charismatic or visionary; instead;

> Even the least charismatic manager or the quietest worker can now be a leader. Such acts of leadership will require 'grungy chores,' will be a 'long hard race,' require 'an act of faith,' and they must not expect a 'cheering crowd.' But the implication is that it will be worth it in the end (or in a future life?). (Turnbull, 2006, p. 266)

The leader who epitomizes the new post 9/11 "anti-hero" model of leadership is former New York Mayor, Rudy Giuliani. While Americans were truly grateful for his leadership in the days following 9/11, we must pay careful attention to the power his story of leadership may represent as it unfolds in the contemporary cultural surround.

What began as an arc of leadership that led from a desire to improve human relations through enlightened and heroic vision (Theory Y) has morphed into a harsher need to secure, control, command, and surveil, which is closely reminiscent of the Cold War Theory X style of leadership.

Additionally, the same strategies Theory X leaders deployed to gain authority and control power through cultivating a culture of fear and maintaining a cult of loyalty without reciprocity that characterized darkside leaders during the Cold War once again reappear in a post-9/11 era, whether those leaders are in government or in industry. The dark side has triumphed, at least temporarily.

How do leaders respond to heightened insecurity and an uncertain environment? They try to control what they can, using the tools they have acquired from our culture. In a media-saturated culture, all of us have learned to become "watchers" who seek unceasing flows of information or data (which is often mistaken for knowledge), often without time for reflection, analysis or critique. We also, following from George Gerbner's celebrated work on cultivation theory, have a generalized "mean world" complex that reinforces our suspicions about others who look, sound, or act different from us. Leaders, in this respect, are no different from the rest of us. They too use the tools they have acquired to make sensible their world and their business environment. Where citizens watch, leaders offer surveillance. Where citizens fear, leaders offer control. Where citizens search for certainty and reassurances, leaders offer (sometimes misguided) certainty and (often false) reassurances.

Rhetorically, the stage has been set for the resurgence of authoritative leaders offering unambiguous and easily consumable but fundamentally ineffective messages and message strategies. As in the Cold War, citizens and workers associate unambiguous statements with truths and tough language with certainty; we associate certainty with improved security. However, unlike the rhetoric that characterized leadership during the Cold War, our post-9/11 leaders rely less on a burden of truth supported by real evidence than on simplistic repetitions of symbolic soundbites (e.g., "September 11 changed everything"; "democracy"; "freedom") clothed in puffed-up patriotic sentiment. Nor is this pattern limited to international political affairs. We find the same pattern in the unethical behavior of Enron executives, World.Com executives, and public appointees entrusted with public safety in times of crisis (Conrad, 2003; Gallagher, Fontenot, & Boyle, 2005; Seeger & Ulmer, 2003;). Who can forget, "You're doing a heck of a job, Brownie!"?[5]

For those of us operating under these discursive conditions there is little cultural or work space available to turn away from the relentless manufacture of fear, incompetence, and the divisive yet enabling rhetoric of an enduring global war. We willingly submit daily to concertive control because this pathetic style of leadership promises something to rely on, something to believe in, even if we know full well that little of it is true (Barker, 1999). We sacrifice our true actualization for pure fantasy. We reduce knowledge creation, management, and *interpretation* to information transfer, we reduce democratic participation to faith in soundbites.

What kind of business leader is called for under these circumstances? What kind of organizational communication is necessary? In times of great uncertainty, what do people rhetorically crave? Judging from recent experience in North America, we can offer the following observations:

- We *want* a leader who "tells it like it is" and accepts responsibility for outcomes (Morris, Schindehutte, Walton & Allen, 2002; Welch, 2001, 2005); we *accept* leaders who are bullies, who tell it like they think that we want it and blame others when it does not work out that way;
- We *want* a leader who takes charge of an organization with machine-

like precision in giving instructions based on the recommendations of informed counsel (Seeger & Ulmer, 2003); we *accept* leaders who claim they are in charge, operate without adequate evidence, and ignore evidence and reasoning that complicates their notions of rightness (Hsu, 2006; National Commission on Terrorist Attacks, 2004).

- We *want* communication that is compassionate, clear, focused, unambiguous, and decisive (Giuliani); we for *settle for* communication that is redundant, simplistic, and divisive.
- We *crave* a leader who has a consistent, value-driven message and command of the language (Johnson, 2001); we *settle for* leaders with a consistent message and an inability to speak off-script.

Overall, then, we desire inspirational leaders who act as heroes and who "express and encourage us to keep the promise of democracy, a promise that respects the truth of who we *are*" (Hyde, 2005, p. 24); but as audience members trying to cope with conditions of heightened uncertainty in a post-9/11 world, we are too often uninterested in hearing the "whole truth about matters of importance" or are "not prepared to deal with the truth and the anxiety" that may accompany it (Hyde, 2005, p. 25). We lessen our expectations for truth and tolerate incursions against our rights in exchange for even temporary states of *perceived* security.

And why do we accept less from our leaders than we know they should be capable of? The easy answer—and maybe the right one—is that we will give up almost all reason in exchange for a sense of, or the hope for, ontological security. Another reason is that in an absence of brightside leaders who embody the higher qualities we associate with earned elite positions in government, industry, and education, we are left with darksiders who traffic in our fear, weakness, and the absence of worthy competitors (Herbert, 2006). Left unchecked, the logical extension of this dark side of leadership is the very embodiment of fundamentalism. We elect and are led by leaders who know their own truth, do not tolerate disloyalty, do not bother with argument or evidence, and who make decisions guided only by their self-interest and faith.

Tracing the "Dark Side" Through History: How Did We Get Here?

As we have argued, our analysis demarcates three distinct historical eras and their attendant leadership theories. However, it is clear that those analytic categories are neither mutually exclusive nor are they entirely distinct. Traces of previous eras reappear, albeit in partial, politicized and sometimes distorted ways, in the present. Moreover, what we take for granted as the "past," particularly in relation to leadership theories, is always already partial. As Cooke (1999) makes clear, the Cold-War assumptions about leadership that we have inherited in management texts have been purged of their "unsavory" elements. Like Maslow, other Cold-War leadership theorists including Kurt Lewin and John Collier, advocated a utopian vision that was not necessarily based on managerial authority, but on a belief in egalitarianism, an opposition to hierarchy, revolutionary change, a commitment to democracy, and the modern progress narrative. The venerated Edgar Schein even admitted his theory of change management was borrowed from the Chinese Communist Party. And yet, the "leftness of the sources of the ideas" of Cold War leadership have been largely written out of extant management theorizing, leaving behind a residue of leadership theory and practice as "technocratic and ideologically neutral" rather than revolutionary and explicitly political (Cooke, 1999, p. 81). Yet, those traces remain, often unspoken, below the surface, and rarely acknowledged.

After the Cold War, leadership theories, stripped from their historical moorings and seemingly devoid of ideological bias, forwarded a version of American business and political leaders as benevolent, truthful and right whose task was to create opportunities for self-actualization or attitude change, not in service of social justice or radical transformation but in service of organizational/managerial goals. Cold-War leadership models assumed that change and leadership efforts should emanate from those located at the top of the (government, organizational, or human need) hierarchy who had the ability to take the long view and the knowledge to make the best decisions. Then those visions were to be taken up by less developed states, organizations and individuals. Those beliefs persist today, in many ways. Indeed, American leaders' opinions about the U.S.'s role in the world have not changed much since the Cold War and the dissolution of the former Soviet Union (Murray & Cowden, 1999).

As the post Cold-War era unfolded, the assumption of change as an inherently good (rather than socio-historical, political, interested process) persisted, but under a new guise and a shifting epistemological foundation. The Cold War utopian vision of America and American political and business leaders and modes of organizing as fundamentally right, good and just required a dystopian counter-point for its truth value. The Soviet Union, Communism and communists served as negative anchors for our positive national narrative and the leadership theories that appropriated that narrative (Cooke, 1999). In the absence of an enemy, a clear and present threat to America and its ideals, the post-Cold War leaders were left without a fixed and stable epistemological foundation.

Thus, in this post Cold-War environment, the truth of a leader's vision became relative to global markets. It was not the best (utopian) vision that won out, but the one that had the catchiest and slickest marketing hook. Where Cold War leaders were known for having heroic (if not elitist) traits, having done something important, or occupying a privileged position in a legitimate social order, the post Cold-War leader was buoyed along by his style, his ability to engage in symbolic management and embody the entrepreneurial spirit of the era (du Gay, 1996). The enterprising leader of the post-Cold War period may have lacked a foundational utopian truth, but traces of the progress narrative of the former era reappear in the form of a (fluid and changing) quest for an ever-better self, organization and even nation.

Yet, devoid of a utopian vision that undergirded the Cold War, the ascension of spin, cultural change, and participatory management processes that emerged in the post-Cold War period were soon appropriated to serve the interests of national/organizational security. In an anxious post-9/11 world, leaders are now able to strategically harness "spin," to deliver messages and enact policies that are ostensibly designed to enhance security, but authoritarian leaders can simultaneously rhetorically and literally refuse to provide evidence for the utility of those messages or policies and, instead, rely on the faith of their followers. Citizens, for example, can not know how and when their phone messages will be recorded, but they can know that it is "good" for them and for national security. When knowledge management is used unethically, it is a potent means for leaders to create increasingly opaque and exclusive communication systems that serve their interests, render others relatively powerless, and enhance their control (Trethewey & Corman, 2001). Moreover, the secrecy that is often the hallmark of control-driven knowledge management systems also prevents the subjects of political and organizational leaders from being able to critically evaluate a leader's message and demands a culture of loyalty (Goodall, 2006c). In the absence of data, citizens and organizational members simply have "faith" in their leaders' pronouncements. To question leaders, to ask for potentially "sensitive" evidence of their claims, is to display a lack of faith, disloyalty and heresy. Leaders who do not tolerate disbelievers may be best understood as fundamentalists, and fundamentalism eliminates the potential for democratic dialogue.[6]

By fundamentalism, we are referring to meaning-making processes that rely on what Eisenberg (2007) refers to as "dominion" narratives. Dominion narratives are characterized by "single meaning" and "the importance of centralized and singular control" (Eisenberg, 2007, p. ix). Fundamentalists, of religious, political, cultural, and corporate stripes, adopt "rigid responses to a perceived gap between one's ideals and the state of the world" (Eisenberg, 2007, p. ix). While fundamentalism is typically used to describe deeply-held religious beliefs, Americans have a tendency to treat ideas about everything from leadership to motherhood to entertainment religiously. As Warner (2005, p. 63) explains:

> It isn't just that we are an extremely religiously observant people. It's that our faith, our inspirational bent, leads us to constantly elevate *everything*—our ideas, our opinions, our tax policies, our diet crazes and stroller choices—to the level of theological doctrine.

When leaders elevate their own perspectives to the level of theological doctrine, when they cling too tightly to the veracity of their own worldviews, perspectives, or dominion narratives, their ability to take the role of the "other" is dangerously compromised. In this way, fundamentalism is the enemy of communication (Eisenberg, 2001) and the very darkest side of leadership. We can trace its trajectory from the shadows of Cold War self-actualization and human motivation theories; witness its development in post-Cold War unbridled optimism for globalization, commodification, and spin; and locate its current manifestation in the failed leadership of certain national leaders, business executives, and disgraced Presidents of educational institutions.

Recovering the Bright Side of Leadership: Embracing Pragmatic Complexity as an Oppositional Narrative

Is the bright side to be recovered? And if so, how? Our view is that there is a bright side available only if we reestablish leadership as a nuanced communicative activity. To do so we must insist that our elected and ascending leaders embody the principles of enlightened self-actualization, empowerment of others, care of the environment, compassion for "others," and ethical actions.

In short, creating the discursive possibilities for the bright side requires that we start by taking "an honest look at ourselves and *the ways we talk to each other*, and the way our leaders talk among themselves and with other leaders throughout the world" (Goodall, 2006b, p. 53).

Enacting the bright side of leadership requires that we develop a fuller understanding of communication than previous leadership theories have allowed. Across the three narrative histories we have explored, communication remains rooted in an "effectiveness" model that assumes that the best message will prevail in the free market place of ideas, that meanings are grounded in personal experience, and that individuals represent their interests simply by giving "voice" to those interests and experiences.

In the effectiveness model, leaders often assume that the "message" is the most important feature of communication, so they overemphasize the pursuit of clarity, consistency, and simplicity. In communication systems that prioritize effectiveness or influence, the overall effect of communication is often to simply reproduce and reinscribe existing systems as leaders and audiences work to "fit" their message into existing interpretive frameworks (Corman, Trethewey & Goodall, 2007).

This model is ill equipped to create and sustain democratic organizations, institutions and societies. Further, it enables the dark side of leadership to continue unfettered by enabling leaders to practice seemingly effective communication as authoritative statements, public relations campaigns, or control strategies.

In place of the "effectiveness" model, we offer a *pragmatic complexity* model of communication (Corman, Trethewey & Goodall, 2007; Singer, 2007). Pragmatic complexity foregrounds "meaning making" as central to the leadership communication process and recognizes that simply repeating messages, narratives or history does little to change existing "darkside" frameworks. Pragmatic complexity is a model of communication that moves away from dominion narratives and towards "engagement narratives" characterized by "multiple meanings, vulnerability, participation and inclusion" (Eisenberg, 2007, p. ix). In an increasingly complex global environment, one of the central challenges leaders face will be communicating with people who hold radically different views and who are passionately connected to them, perhaps in ways that border on fanatical or fundamentalist. Communication in these complex systems must be at once inclusive and preserve important differences. Pragmatic complexity provides an orientation to communication for achieving those goals.

Pragmatic complexity requires leaders to practice and audiences to demand communication that: 1) deemphasizes control and invites complexity and ambiguity as features of contemporary cultures and workplaces that cannot be eliminated, but can be strategically engaged; 2) replaces repetition with variation or experimentation to determine which communication strategies enable social systems to thrive; and, 3) considers disruptive moves capable of "tweaking" existing meaning systems to enable system evolution (Corman, Trethewey, & Goodall, 2007; Singer, 2007). This form of communication is the means through which democracy can be reclaimed and recaptured in moment-to-moment interactions (Deetz, 1995). Here, communication is re-imagined as the process through which foundational meanings—for our selves, our interests, our communities, our nations, our values—are not simply assumed and voiced but are negotiated in and through dialogue. The meanings created through responsive communication are never fixed, but they are provisional, pragmatic, experimental, and fundamentally open to change. In short, those meanings are the cornerstone of democratic cultures (Dewey, 1961).

Conceived thusly, communication requires a new leader who is fundamentally open to ambiguity, or to what we might call ontological insecurity, and leadership theorizing that acknowledges, allows for and embraces uncertainty and complexity. We offer here a way of theorizing leadership that befits our current historical positioning. Pragmatic complexity serves as a both a counternarrative and an appropriate response to leadership practice and theory that slips all too easily into fundamentalism.

Pragmatic complexity assumes three leadership principles. First, leaders should construct and act on *provisional* meanings that are constructed in locally situated and historically informed participatory dialogue. Adopting a provisional approach to meaning should not be understood as abandoning beliefs, values or perspectives. It simply means that leaders cannot adopt fanatical attachments to them (Eisenberg, 2007). Second, brightside leaders test provisional meanings in ongoing social practice.

Leadership is thus *experimental* rather than formulaic, unreflexive, or treated as doctrine. Brightside leaders do not insist on singular interpretations of the world, their environments or their mission but are open to co-constructing new meanings, testing them in the world, and making adjustments based on experiential data. This process is the cornerstone of democratic decision making and the counter-point to fundamentalism. Finally, brightside leaders embrace an *ironic* stance. Irony, as Haraway (1990) defines it, it the "tension of holding incompatible things together because both or all are necessary and true" (p. 190). Irony is an appropriate and effective stance for leaders, particularly in our uncertain times, because it:

> Is a way to keep oneself within a situation that resists resolution in order to act politically without pretending that resolution has come . . . [it is] a vehicle for enabling political action that resists the twin dangers of paralysis (nothing can be done because no final truth can be found) and totalization (there is one

way to do things, the way reflecting the truth that has been found. (Ferguson, 1993, p. 35).

Our analysis of Western leadership theory and practice in the context of historical, cultural and global political struggles, points to a dangerous trend toward the "dark side" of leadership we call fundamentalism. We think that it is increasingly vital for scholars and practitioners to counter this turn by reconsidering, rewriting, and reinvigorating leadership theorizing and practice in ways that enhance individuals', organizations' and nations' democratic and bright human potential. We hope this chapter contributes to that important project.

Notes

1. We agree wholeheartedly with Cooke's (1999) position that "history is inevitably interpretive, with the distinction to be drawn between the 'past' and 'history.' The past comprises an infinity of events, which because of its vastness and pastness cannot ever be fully known . . . History, our knowing of the past, is constructed by identifying some of these events as significant, and, by implication, others as not, and by giving these events particular meanings" (p. 83). This interpretive framework parallels our intention to contribute to a new history of leadership that links themes from selected historical eras to themes prevalent in leadership theorizing.
2. It is important to note here that our analysis of leadership practices and theories since the Cold War focuses on Western, and particularly American, models. However, we believe that the assumptions that those models employ are often the products of larger global struggles. To wit, recent developments in Western leadership theorizing are, in many ways, a response to contemporary concerns around the world about enhancing security in the context and era of the "Global War on Terror" (Turnbull, 2006). It is also the case that Western models of leadership, particularly organizational leadership, have been promulgated, to varying degrees, across the globe as multi-national corporations have assumed greater prominence in many economies and globalization, consumption, and technology have contributed to a "flattened" world (Friedman, 2006).
3. The recent reports alleging that the U. S. National Security Administration has been secretly collecting the phone call records of Americans, with the aid of three of the largest carrier companies, thereby amassing the largest data base ever created. The knowledge management systems that are most problematic for democratic systems are those which are opaque and exclusive. This program is both. The ostensible reason for this program is to enhance national security; however, given that citizens were unaware of this program and have no knowledge of how this

information might be used, its potential for misuse is great (Cauley, 2006).

4. John Rendon epitomizes this idea. Rendon of the Rendon Group consulting firm is a key player in the White House propaganda campaign, a leader in the field of "perception management" and a self-proclaimed "information warrior." He works by monitoring a massive database of knowledge (media stories from across the globe, classified texts, and satellite images) and manipulating responses which often involve disinformation or deception operations that are then fed back to the media to support the President's political objectives (Bamford, 2005). Rendon epitomizes a knowledge management approach to post-9/11 leadership.
5. In the days following the most devastating natural disaster in U. S. history, many Americans and citizens around the globe believed the American government was largely unprepared and woefully unresponsive to the plight of the victims of Hurricane Katrina in New Orleans, Louisiana. And yet, President Bush issued this compliment, which was subsequently repeated in a variety of news outlets, to his friend, supporter and (then) head of the Federal Emergency Management Agency, Michael Brown for his role in the coordinating the emergency response process.
6. When the Bush administration recently dismissed Iranian President Mahmoud Ahmadinejad's letter to President Bush, the first such communiqué from an Iranian leader to a US leader in a quarter century, as merely "meandering screed," they effectively eliminated the possibility of engaging in a politically and symbolically significant conversation. (Goodall et al., Chapter 5, this volume) and may have further entrenched anti-American sentiment across the globe (Kohut & Stokes, 2006).

References

Ashcraft, K. L., & Mumby, D. K. (2004). *Reworking gender: A feminist communicology of organization.* Thousand Oaks, CA: Sage.

Ackman, D. (2002, September). The 20 most influential business books. *Forbes.com.* Retrieved May 15, 2007 from http://www.forbes.com/2002/09/30/0930booksintro_2.html

Awad, W. M., & Ghaziri, H. M. (2003). *Knowledge management.* Upper Saddle River, NJ: Prentice Hall.

Bamford, J. (2005, November 17). The man who sold the war. *Rolling Stone*. Retrieved May 18, 2006 from *http://www.rollingstone.com/politics/story/8798997/the_man_who_sold_the_war.*

Barker, J. R. (1999). *The discipline of teamwork: Participation and concertive control.* Thousand Oaks, CA: Sage.

Barley, S. R., & Kunda, G. (1992). Design and devotion: Surges of rational and normative ideologies of control in managerial discourse. *Administrative Science Quarterly,*

37, 363–400.
Bennis, W. (1994). *On becoming a leader*. New York: Perseus Books.
Biggart, N. W. (1983). Rationality, meaning, and self-management: Success manuals, 1950–1980. *Social Problems, 30*, 298–311
Blake, R., & Mouton, J. (1964). *The managerial grid*. Houston: Gulf.
Blase, J., & Blase, J. (2002). The dark side of leadership: Teacher perspectives of principal mistreatment. *Education Administration Quarterly, 5*, 671–727.
Bolman, L. G., & Deal, T. E. (1991). *Reframing organizations: Artistry, choice and leadership*. San Francisco: Jossey-Bass.
Burns, J. M. (2005). Leadership. *Leadership*, 1, 11–12.
Cauley, L. (2006, May 10). NSA has massive data base of Americans' phone calls. *USA Today*. Retrieved May 16, 2006 from http://www.usatoday.com/news/washington/2006–05–10-nsa_x.htm.
Collins, J., & Porras, J. I. (1994). *Built to last: Successful habits of visionary companies*. New York: Harper Collins.
Collinson, D. (2005). Dialectics as leadership. *Human Relations, 58*, 1419–1443.
Collinson, D., & Grint, K. (2005). Editorial: The leadership agenda. *Leadership, 1*, 5–9.
Conrad, C. (2003). Setting the state: Introduction to the special issue on the 'corporate meltdown'. *Management Communication Quarterly, 17*, 5–19.
Cooke, B. (1999). Writing the left out of management theory: The histography of management change. *Organization, 6*, 81–105.
Cooke, B., Mills, A. J., & Kelley, E. S. (2005). Situating Maslow in Cold War America: A recontextualization of management theory. *Group & Organization Management, 30*, 129–152.
Corman, S., Trethewey, A., & Goodall, H. L. Jr. (2007). A 21st century model for communication in the global war of ideas: From simplistic influence to pragmatic complexity. Retrieved May 1, 2007 from *http://comops.org/publications/CSC_report_0701-pragmatic_complexity.pdf*
Covey, S. R. (1992). *The seven habits of highly effective people*. New York: Simon & Schuster.
Czarniawska-Joerges, B., & Wolff, R. (1991). Leaders, managers and entrepreneurs on and off the organizational stage. *Organization Studies, 12*, 529–546.
Deal, T., & Kennedy, A. (1982). *Corporate cultures: The rites and rituals of corporate life*. Reading, MA: Addison-Wesley.
Deetz, S. (1995). *Transforming communication, transforming business: Building responsible and responsive workplaces*. Cresskill, NJ: Hampton Press.
Dewey, J. (1961) *Democracy and education*. New York: Macmillan.
Dubner, S., & Levitt, S. J. (2005). *Freakonomics: A rogue economist explores the hidden side of everything*. New York: William Morrow & Co.
du Gay, P. (Ed.), (1997). *Production of culture/cultures of production*. London: Sage.

du Gay, P. (1996). *Consumption and identity at work.* London: Sage.
Dye, K., Mills, A. J., Weatherbee, F. C. (2005). Maslow: Man interrupted: Reading management theory in contexts. *Management Decision, 43*, 1375–1395.
Eisenberg, E. M. (2007). *Strategic ambiguities: Essays on communication, organization, and identity.* Thousand Oaks, CA: Sage.
Eisenberg, E.M. (2001). Building a mystery: Communication and the development of identity. *Journal of Communication*, 51(3), 534–552.
Eisenberg, E.M., Goodall, H.L., & Trethewey, A. (2007). *Organizational communication: Balancing creativity and constraint,* 5th ed. New York: Bedford/St. Martin's.
Fiedler, F. (1967). *A theory of leadership effectiveness.* New York: McGraw-Hill.
Ferguson, K. E. (1993). *The man question: Visions of subjectivity in feminist theory.* Berkeley, CA: University of California Press.
Foucault, M. (1979). *Discipline and punish: The birth of the prison.* (trans A. Sheridan). New York: Vintage.
Friedman, T. (2006). *The world is flat: A brief history of the twenty-first century.* New York: Farrar, Straus and Giroux.
Gaddis, J. L. (2005). *The cold war: A new history.* New York: Penguin.
Gallagher, A. H., Fontenot, M., & Boyle, K. (2005). Communicating in times of crises: An analysis of news releases from the federal government before, during and after hurricanes Katrina and Rita. *Public Relations Review, 33*, 217–219.
Gerbner, G., Gross, L., Morgan, M., & Signorielli, N. (1986). Living with television: The dynamics of the cultivation process. In J. Bryant & D. Zillmann (Eds.), *Perspectives on media effects* (pp. 17–40). Hillsdale, NJ: Erlbaum.
Gladwell, M. (2007). *Blink: The power of thinking without thinking.* Boston: Back Bay Books.
Gladwell, M. (2002). *The tipping point: How little things can make a big difference.* Boston: Back Bay Books.
Goodall, H. L., Jr. (2006a). *A Need to know: The clandestine history of a CIA family.* Walnut Creek, CA: Left Coast Press.
Goodall, H. L., Jr. (2006b) Why we must win the war on terror: Communication, narrative and the future of national security. *Qualitative Inquiry, 12*, 30–59.
Goodall, H. L., Jr. (2006c, May 19). My family secret. *Times Higher Education Literary Supplement*, Retrieved May 19, 2006 from
Grant, J. D., & Mills, A. J. (2006). The quiet Americans: Formative context, the Academy of Management leadership, and the management textbook, 1936–1960. *Management and Organizational History, 1*, 201–224.
Giuliani, R. & Kurson, K. (2002) *Leadership.* New York: Miramax.
Haraway, D. (1990). A manifesto for cyborgs: Science, technology and socialist feminism. In L. J. Nicholson (Ed.), *Feminism/postmodernism.* (pp. 190–233). New York: Routledge.
Herbert, B. (2006, May 15). America the fearful. *New York Times,* Op. Ed. Page.

Hersey, P., & Blanchard, K.H. (1969) *Management of organizational behavior*. Englewood Cliffs, NJ: Prentice-Hall.

Holmberg, I., & Strannegard, L. (2005). Leadership voices: The ideology of the 'new economy'. *Leadership, 1*, 353–374.

Hsu, S. S. (February 12, 2006). Katrina report spreads blame. *Washington Post*, A01.

Iverson, J., & McPhee, R. D. (2003). Knowledge management in communities of practice: Being true to the communicative character of knowledge. *Management Communication Quarterly, 16*, 259–266.

Jacques, R. (1996). *Manufacturing the employee: Management knowledge from the 19th to the 21st centures*. London: Sage.

Jansen, S. C. (2005). Foreign policy, public diplomacy, and public relations: Selling America to the world. In. L. Artz & Y. R. Kamalipour (Eds.), *Bring 'em on: Media and politics in the Iraq War.* Lanham, MD: Rowman & Littlefield.

Johnson, C. (2001). *Meeting the ethical challenges of leadership: Casting light or shadow*. Thousand Oaks, CA: Sage.

Johnson, S., & Blanchard, K. (1998). *Who moved my cheese? An amazing way to deal with change in your work and in your life.* New York: Putnam.

Judt, T. (2005). *Postwar: A history of Europe since 1945.* New York: Penguin.

Knights, D., & Willmott, H. (1999). *Management lives: Power and identity in work organizations*. London: Sage.

Kohut, A., & Stokes, B. (2006). *America against the world: How we are different and why we are disliked.* New York: Times Books.

Kunda, G., & Souday, G. (2004). New designs: Design and devotion revisited. In S. Ackroyd, R. Batt, P. Thompson & P. S. Tolbert (Eds). *The Oxford handbook of work and organization* (pp. 184–206). Oxford: Oxford University Press.

Lair, D. J., Sullivan, K., & Cheney, G. (2005). Marketization and the recasting of the professional self: The rhetoric and ethics of personal branding. *Management Communication Quarterly, 18*, 307–343.

Levitt, S. J., & Dubner, S. (2005). *Freakonomics: A rogue economist explores the hidden side of everything.* New York: William & Morrow.

Lutgen-Sandvik, P. (2003). The cycle of employee emotional abuse: Generation and regeneration of workplace mistreatment. *Management Communication Quarterly, 16*, 471–501.

Maslow, A. (1965). *Eupsychian management*. Homewood, IL: Irwin.

McGregor, D. (1960). *The human side of enterprise*. New York: McGraw-Hill.

McGregor, D., Bennis, W. G., Schein, E. H., and McGregor, C., Eds. (1968). *Leadership and motivation : Essays of Douglas McGregor*. Cambridge, MA: MIT Press.

Morris, M. H., Schindehutte, K., Walton, J., & Allen, J. (2002). The ethical context of entrepreneurship: Proposing and testing a developmental framework. *Journal of Business Ethics, 40*, 331–361.

Mumby, D. K. (1988). *Communication and power in organizations: discourse, ideology,*

and domination. Norwood, NJ: Ablex.
Murray, S. K. & Cowden, J. A. (1999). The role of 'enemy images' and ideology in elite belief systems. *International Studies Quarterly, 43*, 455–481.
National Commission on Terrorist Attacks (2004). *The 9/11 Commission Report*. New York: W.W. Norton & Company.
Parrett, B. (2007). *The sentinel CEO: Perspectives on security, risk, and leadership in a post-9/11 world*. New York: Wiley
Peters, T. (1999). *The brand you 50: Fifty ways to transform yourself from an 'employee' into a brand that shouts distinction, commitment, and passion!* New York: Knopf.
Peters, T. J., & Waterman, R. H. (1982). *In search of excellence: Lessons from America's best run companies*. New York: Harper Collins.
Pfeffer, J., & Sutton, R. (2006). *Hard facts, dangerous half-truth and total nonsense: Profiting from evidenced-based management*. Cambridge: Harvard Business School Press.
Porter, M. E. (1998a). *Competitive advantage: Creating and sustaining superior performance.* New York: Free Press.
Porter, M. E. (1998b). *The competitive advantage of nations*. New York: Free Press.
Quinn, R. E. (2005). Moments of greatness: Entering the fundamental state of leadership. *Harvard Business Review, 83,* 74–83.
Redding, W. C. (1985). Stumbling toward identity: The emergence of organizational communication as a field of study. In R. D. McPhee and P. K. Tompkins (Eds.) *Organizational communication: Traditional themes and new directions* (pp. 15–54). Beverly Hills, CA: Sage.
Robbins, A. (1992). *Awaken the giant within: How to take immediate control of your mental, emotional, physical and financial destiny!* New York: The Free Press.
Runté, M., & Mills, A. J. (2006). Cold war, chilly climate: Exploring the roots of gendered discourse in organization and management theory. *Human Relations, 59*, 695–720.
Seeger, M. W., & Ulmer, R. R. (2003). Explaining Enron: Communication and responsible leadership. *Management Communication Quarterly, 17*, 58–84.
Senge, P. (1991). *The fifth discipline: The art and practice of the learning organization*. New York: Doubleday/Currency.
Singer, P.W. (2007, May 16). America's new leadership: Reboot and restart. Retrieved May 17, 2007 from *http://www.washingtonpost.com/wp-dyn/content/article/2007/05/15/AR2007051500995.html*
Taylor, M. L. (2005). *Religion, politics and the Christian right: Post-9/11 powers in the American empire.* Minneapolis, MN: Augsburg Fortress Press.
Townley, B. (1994). *Reframing human resource management: Power, ethics and the subject at work*. London: Sage.
Trethewey, A., & Corman, S. (2001). Anticipating k-commerce: E-Commerce, knowledge management, and organizational communication. *Management Communica-*

tion Quarterly, 14, 619–628.

Turnbull, S. (2006). Post-millennial leadership refrains: Artists, performers and anti-heroes. *Leadership, 2*, 257–269.

Warner, J. (2005). *Perfect madness: Motherhood in the age of anxiety*. New York: Riverhead Books.

Welch, J. (with) Welch, S. (2005). *Winning*. New York: Collins.

Welch, J. (with) Byrnc, J. A. (2001). *Straight from the gut*. New York: Warner.

Wheatley, M. (1992). *Leadership and the new science*. New York: Wiley.

White, R. K., & and Lippitt, R. (1960). *Autocracy and democracy: Experimental inquiry*. New York: Harper & Brothers.

Whyte, W. (1956). *The organization man*. New York: Simon & Schuster.

Zorn, T. E, Page, D, J., & Cheney, G. (2000). Nuts about change: Multiple perspectives on change-oriented communication in a public-sector organization, *Management Communication Quarterly, 13*, 515–566.

PART II

Case Studies on Communication, Terrorism, & National Security

CHAPTER FOUR

Communication and Media Strategy in the Islamist War of Ideas[1]

STEVEN R. CORMAN AND JILL S. SCHIEFELBEIN

WITH CONTRIBUTIONS FROM: KRIS ACHESON, IAN DERK, BUD GOODALL, AARON HESS, Z. S. JUSTUS, KELLY MCDONALD, ROBERT MCPHEE, CHRISTINA SMITH, ANGELA TRETHEWEY, MARK WOODWARD

Introduction

Recent controversies surrounding U.S. efforts to influence media in Iraq and the Middle East signal increasing interest in a war of ideas that is part of the conflict between the West and the worldwide Islamist movement.[1] Clear thinking about this issue requires an understanding of how the Islamists struggle for hearts and minds. Yet many people are under-informed about the nature and extent of Islamist strategy regarding communication and the media. The purpose of this chapter is to piece together this strategy from texts captured during operations in Afghanistan and Iraq, translated statements from Islamist leaders, as well as other open-source documents, such as speeches and website material.

These texts reveal three strategic goals for communication and media in Islamist operations. First, they must **legitimate** their movement by establishing its social and religious viability while engaging in violent acts that on their face seem to violate the norms of civilized society and the tenets of Islam. This is perhaps the biggest ongoing communication challenge they face. Second, they aim to **propagate** their movement by spreading messages to sympathetic audiences in areas where they want to expand. This prepares the way for political efforts that precede establishing actual operations. Third, they seek to **intimidate** their opponents. This applies not only to existing enemies but to sympathizers in the Muslim world who might think of turning against them.

Islamists pursue these strategies using sophisticated, modern methods of communication and public relations. They segment audiences and adapt their message to the audience, apply some of the same PR techniques used by large corporations, conduct disinformation campaigns, and coordinate communication with operations. They do this using a variety of sophisticated means, including traditional mass media and new media channels. This paper places particular emphasis on new media channels, especially the Internet, to understand the implications of a "virtual jihad."

Six recommendations are generated from our analysis: (1) adopt a long-term strategy of improving our credibility with Muslim audiences, (2) degrade Islamists' ability to execute their communication and media strategy, (3) identify and draw attention to Islamist actions and outcomes that contradict Islam, (4) deconstruct Islamist concepts of history and audience, (5) redouble efforts to engage Islamist new media campaigns, and, (6) make better use of sympathetic members of the American Muslim community.

Background

We developed this chapter in response to recent controversies about efforts by the United States to influence foreign media coverage of Islamist activities in Iraq and the Middle East. We join this debate not by taking a position on one side or the other, but instead by focusing on the "flip side" of the issue: The efforts by Islamists to influence media coverage of their activities. Whether a particular operation on the U.S. side is justified is a legitimate matter of argument. But while we deliberate such issues, the Islamists are busy executing a communication and media strategy of their own. It is designed to spread their ideas, proliferate their movement, and intimidate their enemies through traditional and new media. It seems beyond question that the United States should follow and resist these efforts, even if the proper methods for doing this remain an open question.

The goal of this chapter, therefore, is to describe the strategies and principles underlying Islamist communication and media efforts. The role of communication practices in promoting jihad "has been systematically undervalued" (International Crisis Group, 2006, p. 4) in discussion of how best to resist the movement. Osama bin Laden, in a letter to Emir Al-Momineen, states: "It is obvious that the media war in this century is one of the strongest methods; in fact, its ratio may reach 90% of the total preparation for the battles" (AFGP-2002–600321). Yet we find that many otherwise well-informed people are surprised to learn that this "battlefield" even exists. They do not realize that Islamists have an explicit communication and public relations strategy, that they execute this strategy in a sophisticated manner that makes use of modern tools and techniques, and that they are rapidly assimilating new media into their repertoire in hopes of establishing a worldwide virtual jihad movement.

Just as surprising as the extent of Islamist communication strategy is the ease with which it is understood. It does not take sophisticated analysis and secret spy operations to understand their plans; we need only to listen to what they are saying. Accordingly, this review relies primarily on quotes from recently de-classified captured documents and other open source reports. The main documents consist of 28 texts captured in operations against al Qaeda and the Taliban in Afghanistan and "range from single page letters to 70+ page excerpts from larger Islamist texts, and were authored both before and after September 11, 2001" according to the Combating Terrorism Center (2006a) report in which they were released. They are cited with the letters "AFGP" to correspond with the numbers used in that report.

One hazard of working with documents like this is that we may be analyzing an organization of the past. There is broad agreement in the counter-terrorism community that the al Qaeda of 9/11—an organized army of Islamist special operations personnel complete with training facilities—is a thing of the past (Combating Terrorism Center, 2006a; Hoffman, 2004). Today it is more of an *ideal* or social movement that is replicated by relatively disconnected groups (as in the cases of the Madrid and London bombers) than a network of cells controlled by a "mother ship." This is the reason we use the more general term *Islamist*, referring to organized Salafi extremists, rather than specific present or past groups like al Qaeda.

"More than half of this battle is taking place in the battlefield of the media. We are in a media battle in a race for the hearts and minds of Muslims."

Abu Musab al-Zarqawi

Analyzing older documents in light of this change is not too much of a worry for the present analysis. We are not interested so much in specific tactics, but rather in the Islamist *concept of operations* with respect to communication and the media. As the movement metastasizes, we can expect this to spread right along with the other ideological machinery. Indeed there is recent evidence that contemporary Islamists show unabated zeal for media operations. U.S. Secretary of Defense Donald Rumsfeld recently quoted al Qaeda in Iraq (AQI) leader Abu Musab al-Zarqawi as saying, "More than half of this battle is taking place in the battlefield of the media. We are in a media battle in a race for the hearts and minds of Muslims" (Rumsfeld, 2006).

We also include as a point of contemporary reference three more recently released documents. The U.S. military recently conducted an operation against al AQI, capturing a Islamist who was carrying documents. Three of these were subsequently released for distribution (Combating Terrorism Center, 2006b). We include references to these letters in particular as a way of demonstrating that some of the tactics described in this chapter are alive and well in contemporary Islamist practice. They are cited with "IZ" followed by numbers. Other sources include translated statements of Islamist leaders, as well as speeches, documents and web site material, all available in the open source literature.

In the sections below we describe the Islamist thinking about communication and media, and some of the approaches they have for putting their plans into action. The first section describes overall Islamist strategy and then focuses on communication and media strategy in general, showing that it is designed to legitimate, propagate, and intimidate. The second section shows how the Islamists follow modern principles and fundamental practices in executing their communication and media efforts. The third shows that Islamists view new media—especially Internet-based communication and information technologies—as a platform for global operations and virtual jihad. We conclude by arguing that our ability to directly interfere with Islamist communication efforts is limited, but that within these constraints several strategic opportunities are apparent.

Communication and Media in Islamist Strategy

Our first point is that Islamists place great importance on communication and media as elements of their overall strategy. A *strategy* is a plan that leads to a goal, and is distinct from tactics, which have to do with using resources to execute plans. The goal of the Islamists is a matter of some dispute and depends on the time horizon in question. There is general agreement that in the short run they want to drive those they see as invaders (e.g., the United States and its allies) from the Arabian Peninsula. This goal has been regularly professed by

Osama bin Laden, most recently in his offer of a "truce" to the United States (Bin Laden Tape, 2006). In Iraq, where the Islamists are currently most active, "the armed opposition's avowed objectives have thus been reduced to a primary, unifying goal: ridding Iraq of the foreign occupier. Beyond that, all is vague" (International Crisis Group, 2006, p. 11).

The lack of agreement on longer term goals should not be mistaken for a lack of ideas. One stated intermediate goal of al Qaeda is the toppling of "apostate regimes" in the Middle East. This would lead to restoration of an Islamic Caliphate of the kind that ruled the region in the Middle Ages. Having established this Caliphate, an even longer-term goal is to expand its influence and rule as far as possible, if not worldwide. One of the goals often expressed in Islamist texts is the fulfillment of a desire to be governed by devout followers of Islam. In "A Response to Accusations Against Sheikh Albani," Ayman Zawahiri explicitly voiced this goal, demanding that the infidel, turncoat, and backsliding leaders who currently do not follow Islamic Sharia begin to govern by the revelations of Allah, and urging readers to kill them if they will not. He prioritizes this goal of a devout Islamic Caliphate above all others, including attacks on Western Imperialists and the retaking of Palestine from Israel (AFGP-2002- 601041).

Our goal here is not to dwell on overall Islamist strategy but to show the importance they place on media strategies that help facilitate their overall goals. One sign of importance is the incorporation of communication and media functions in Islamist organizational structures. For example, the formal structure of al Qaeda includes military, political and information committees. The *military committee* is responsible for operations. The *political committee* interacts with the wider jihad movement, establishes political relationships, and maintains relations with the government of the host country. The *information committee* is responsible for the "means of communications setup in all categories of Islamic people, taking great pains in making it aware of its enemies' plans, aspiring to concentrate all of the scientific, legal, and Jihad capabilities in the first level in order to obstruct one line in front of the alliance of the infidel and the ugly ones." (AFGP-2002–000078). Our examination of documents describing the activities of these committees reveals three main strategic goals of legitimating, propagating, and intimidating. We describe these next.

Legitimating

Islamists see themselves as "outsiders" and they use violence to achieve their overall goals. Violent methods inevitably harm innocent people, so there is a built-in drag on the organization's legitimacy. Worse for the Islamists, their ideology is also heavily rooted in Islam, which on its face appears to prohibit the kinds of violent methods they use. So not only is there a huge need to establish and maintain social and religious legitimacy, there is also a rich source of contradictions to complicate this effort. Islamists recognize this and have been concerned with legitimating their efforts "from the start" (International Crisis Group, 2006, p. 8). This is perhaps the biggest communication challenge they face, and accordingly it is given much attention in their writings.

Social legitimation means having the communities in which they operate know their story, share their goals, and accept and support their efforts. This is a central focus of "A Memo to Sheikh Abu Abdullah" (AFGP-2002–003251). In it the author, Abu Huthayfa, laments that "most of the people inside [the country] are unaware of the great effort the mujahidin made against the American forces" in Somalia. He urges use of the media to spread these stories because "publicizing those events will motivate and encourage the nation, breaks the barrier of fear and it gives a live and actual example of the recent experiment in which the mujahidin succeeded in achieving the target and driving the enemy away." Spreading word of the heroism of the mujahidin will "arouse the nation and affirm the movement's credibility." Another document notes how previous Islamist efforts have failed on this account:

> The mujahideen failed to define their identity, their intentions and motivations; such an explanation was and still is the main pillar for attracting the masses and mobilizing the base members on an intellectual and ideological level to partake in this dangerous work (i.e., Jihad). (AFGP-2002–600080)

Religious legitimation means having the jihadis' efforts be seen as acceptable under the religious tenets of Islam. As noted already, on its face Islam would seem to ban their violent practices. For example, in one passage the Quran condemns the killing of Muslim children:

> They are lost indeed who kill their children foolishly without knowledge, and forbid what Allah has given to them forging a lie against Allah; they have indeed gone astray, and they are not the followers of the right course. (Quran 6:140)

Islamist suicide bombers routinely kill children "foolishly without knowledge." Have they not, then, "gone astray?" Interpretation of the Quran is a complex matter of competing arguments by scholars. Thus Islamists spend considerable rhetorical effort promoting their interpretations in this discourse and refuting or preempting criticism by outsiders. For example, in Iraq

> The more active groups now appeal to the same Koranic passages, and tend to interpret current events through the prism of the Crusades (of which U.S. imperialism is seen as the latest manifestation), and invoke mythical/religious events and people (the battle of Hittin in the early age of Islam; the heroic figures of Saladin, liberator of Jerusalem, and al-Qa'qa'; the early Muslim fighters, and so forth). (International Crisis Group, 2006, p. 10)

They also try to mitigate the need for such interpretive defenses by letting others take the blame:

> So the brothers raided his house in the middle of the night wearing the national guards uniforms and driving similar cars, they took him and killed him thank god. The next morning his households and neighbors started talking to the tribe saying it was the national guards, they added that they heard some of them speaking English, meaning that the Americans are the ones who took Abu Dhari . . . no one claimed responsibility for this killing, thank god. (IZ-060316–01).

Propagating

Brachman (in press) points out that "in recent decades, jihadi ideologues have focused significant energy on crafting and implementing an aggressive, historically informed and universally applicable strategy to take over the world" (p. 6). Clearly then the Islamists do not want to merely legitimate their actions but also to spread their movement to other areas and groups, linking their efforts to a wider struggle involving Muslims everywhere. This goal is discussed at length in a document entitled "Interior Organization" (AFGP-2002–000080), a bylaws-like tract establishing the structure of al Qaeda. In the opening section on their principles of "general politics" we find:

Islamists want to link their movement to a wider struggle involving Muslims everywhere.

> 5. Our relation with Islamic movements and groups and workers of Islam [TN—the term "friendly Jihad" is crossed out here and replaced with "workers of Islam"] is one of cooperation towards righteousness and strength, with continuing attempts towards merger and unity.
>
> 6. Our relation with non-Jihad Islamic groups is one of love and friendship and advice, and bringing out the good in them and correcting their mistakes if the situation requires it.

> 9. Eliminate regionalism and tribalism. We struggle in any place in Islamic countries if the situation requires it and our capabilities allow it.
>
> 10. The concern over the role of the Muslim people in the Jihad. And struggle to agitate (them) so that they will be in the rank of al-Jihad because they are fuel for the battle.
>
> (AFGP-2002–000080; items 7 and 8 were "crossed out" in the original document)

In executing these principles the Islamists make concerted efforts to develop media contacts outside their immediate sphere of influence because they view the media as the vehicle that will make their message heard world-over (AFGP-Book by Mustafa Hamid). For example, in a letter to Mohammed Omar, Osama bin Laden recounts many requests for interviews from the press, and suggests that "this is a good opportunity to make Muslims aware of what is taking place over [in] the land of two Holy Mosques as well as of what is happening here in Afghanistan" (AFGP-2002–600321). Islamists view this political communication as very important because it paves the way for establishing bases of operations in other countries, allowing the Islamistst movement to take a step forward in achieving their overall goals (AFGP-2002–600113).

Intimidating

Notwithstanding the division into the military, political, and information committees described above, the Islamist concept of organization reflects tight integration of the three functions. Information activities spread the message to Muslims worldwide, which provides the basis for political activities that form relationships needed to spread the movement. These relationships help establish bases of operations. The ultimate goal is the fight, and everything else is simply part of a "supply chain" that prepares for battle. In this sense, then, legitimating and propagating the movement are themselves part of the intimidation strategy.

Yet Islamists also seem to make more direct use of communication and media in an attempt to scare and intimidate their opponents, a function closer to what we traditionally think of as propaganda. One consistent theme is *putting the enemy on notice* that there can be no room for compromise. The bylaws-like document described above states this in unequivocal terms:

> Our position on dictators of the earth [TN—hegemons?] and secular groups and others that resemble them is that there are innocents among them and un-

> believers among them and that there will be continuing enmity until everyone believes in Allah. We will not meet them half way and there will be no room for dialogue with them or flattery towards them. (AFGP-2002–000080)

They express especially harsh intentions toward the Muslim rulers whom they view as apostate. They have gone so far as to send "open letters" threatening and taunting these rulers (AFGP-2002–000103), presumably in an effort to intimidate them.

Islamists also attempt to scare and intimidate the "far enemy," especially the leaders and people of the United States. In his "Letter of Threat to the Americans" in 2002 (AFGP-2002–001120) Abu Abullah Al Kuwaiti issues a number of statements and threats of this kind. As Table 1 shows, his message contains six themes that bear a remarkable similarity to a statement issued by Osama bin Laden in January 2006 (Bin Laden Tape, 2006). Given the sophistication of Islamist communication and media practices in general, it would not be surprising to learn that these themes reflect "talking points" recorded in information committee documents somewhere.

Table 1: Comparison of themes used by al Kuwaiti and bin Laden in media statements.

Theme	al Kuwaiti 2002	bin Laden 2006
We seek direct dialog.	The statement/letter should be directed to the American people.	My message to you is about the war in Iraq and Afghanistan and the way to end it.
We didn't want to fight you, but you have made us do it.	There is no animosity between us. You involved yourselves [Europe and US] in this battle. The war is between us and the Jews. You interfered in our countries and influenced our governments to strike against Muslims	Based on the above, we see that Bush's argument is false. However, the argument that he avoided, which is the substance of the results of opinion polls on withdrawing the troops, is that it is better not to fight the Muslims on their land and for the enemy not to fight us on our land.

(Table 1 continued on following page.)

Theme	al Kuwaiti 2002	bin Laden 2006
We easily penetrate your security measures.	The groups that are present in Europe and the US are above suspicion. We obtain our intelligence information from your government and intelligence agencies.	On the other hand, the mujahideen, praise be to God, have managed to breach all the security measures adopted by the unjust nations of the coalition time and again.
Your leaders are inept and/or corrupt.	Isn't it time to end American arrogance and begin listening to your people before you experience more devastating disasters?	Bush said: It is better to fight them on their ground than the enemy fighting us on our ground. In my response to these fallacies, I say: The war in Iraq is raging, and the operations in Afghanistan are on the rise in our favour, praise be to God.
Things are getting worse for you and better for us.	I am pleased to inform you the billions you have spent fighting us so far have resulted in killing a small number of us.	Praise be to God, our conditions are always improving and becoming better, while your conditions are to the contrary of this.
We continue to plan attacks against you.	We warn you that our war against you has not ended, but its effects will increase.	Operations are under preparation, and you will see them on your own ground once they are finished, God willing.

Messages in these six themes contribute to the Islamist media strategy of legitimating (things are getting worse for you and better for us), propagating (we continue to plan attacks against you) and intimidating (we easily penetrate your security measures). Methods for disseminating these themes and other Islamists media messages are discussed next.

Methods and Planning

Our second point is that Islamists use sophisticated methods to plan and execute their communication and media operations. By "sophisticated" we mean that they use principles that align with modern methods of communication and public relations. The scope of Islamist PR may not match the multi-million dollar efforts of public corporations, but the point is that their messages are engineered using similar principles and probably gain effectiveness as a result. The documents reviewed in this study revealed evidence of audience segmentation and adaptation, use of tools of the trade, use of disinformation, and coordination of media with operations.

Audience Segmentation and Adaptation

Understanding the Audience

The most fundamental rule of any communication effort is to understand the audience. Islamist communicators appear to have taken this principle to heart by regularly applying conceptual distinctions when discussing their enemies and/or their communication efforts against them. One distinction in evidence is between people inside and outside the movement. For the *insiders*, the communication problems have to do with controlling an amorphous, distributed, secretive organization and orienting everyone to common objectives. This has been discussed in a recent report (Combating Terrorism Center, 2006a) so we will not dwell on it here. But the distinction is still relevant because the insiders are responsible for executing the legitimation and propagation strategies described above. In "Lessons learned from the armed Jihad ordeal in Syria" the unknown writer says that one of the failures was:

> 6th: Week [sic] public relations campaign both inside and out:
>
> We talked previously about the failure of the mujahideen on the inside to propagate their vision, goals and slogans in a clear way easy enough for the people to comprehend and support. They did not have a planned communicable public relations campaign capable of mobilizing their base, backers or supporters. They only issued few ineffective communiqués. (AFGP-2002–600080)

The *outsiders* are divided into categories. A basic distinction is between the good guys and the bad guys. *Good guys* are the "backers, and supporters" of the previous quote, the good Muslims who provide emotional and/or material support for the Islamists' efforts, and Muslims who could potentially be brought into that fold. Islamists have almost parental attitudes toward this group, viewing their relationship as "one of love and friendship and advice, and bringing out the good in them and correcting their mistakes" (AFGP-2002–000080).

The *bad guys* include everyone who is not a good guy, and there are two varieties of these. The *apostates* are "fallen" Muslims who are the most immediate objects of Islamist scorn (and operations), especially if they are rulers of countries. Sometimes described as the "near enemy," this is the group against which the Islamists define their social identity, and who seem to be the main targets of their short term goals. The *unbelievers* are the foreigners, especially those in the West, who are sometimes referred to as the "far enemy." These outsiders are problematic in the short term because they meddle in the affairs of the Arabian Peninsula. They also figure in the long-term dreams of a worldwide Caliphate.

Islamist Plan

1. **Depose the apostates which allows . . .**
2. **Formation of the Caliphate which puts the Islamists in a position of strength from which to . . .**
3. **Attack the Jews which will enable the Islamists to . . .**
4. **Conquer the West.**

Two other groups of outsiders appear in Islamist writings. The *troublemakers* include people like members of "the deposed regime, the tribal cliques, the hired fighters, and the standard criminals" (AFGP-Thoughts about the security of principal squads). They are viewed as enemies of the movement, but less serious enemies than the bad guys, and possibly redeemable. The *Jews* are the penultimate objects of the Islamists' wrath, yet the level of contempt expressed toward them in our sample of documents seems no different than that heaped on the apostates. Indeed the plan seems to be to depose the apostates, which will allow formation of the Caliphate, which will then be in a position of strength from which to attack the Jews, which can then move on to the West. Though we found no explicit message planning for these specific groups, they seem to fill in the cracks between the good guys and bad guys to complete the Islamist audience concept.

Figure 1. Islamist Audience Concept

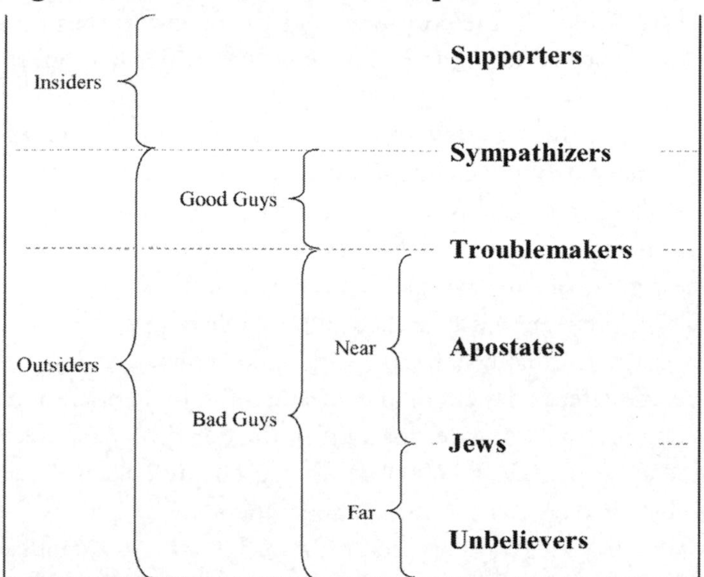

Adapting the Message and Medium

The second most fundamental rule of any communication effort is to adapt the message and the medium to the audience. Here again, there is evidence that the Islamists are following the rule. Sympathizers, as well as potential sympathizers in the lands of the near and far enemies, are primarily targeted with social and religious legitimation messages. Personal communication and other face-to-face methods like speeches and sermons have always been preferred for this task, but as we discuss below the Internet is coming into the mix, especially as a way of cultivating sympathizers in far-away places.

> **"Jihad radio stations operating in Yemen and Somalia will have a more powerful effect on the enemy than nuclear bombs."**
>
> **Hassan al-Tajiki**

Increasingly, there is evidence that good guys are also being targeted with intimidation messages. In one of the AQI documents the unknown author concludes that "we should make the public and the ignorant know clearly our policy fighting the renegades, so they won't side with them for their fight against the Mujahidin" (IZ-060316–01). The same document also shows that the troublemakers category is something of a moving target that slides toward those who do good things for the people. In describing its "restrictions on beheading the heads of the Islamic Party members" AQI says its policy is to "first kill the corrupted and the ones with negative effects, before the ones who try to guide people, and have a positive effect" (IZ-060316–01).

The bad guys are, naturally, targeted to a greater extent with intimidation messages. For the near enemy local media come into the mix because, as discussed above, the Islamists view this kind of coverage as a good means of criticizing existing leaders in preparation for political operations that build relationships with new members and sympathizers. For example, in "Five Letters to the Africa Corps" Hassan al-Tajiki advocates establishing pirate radio stations and predicts that "jihad radio stations operating in Yemen and Somalia will have a more powerful effect on them than nuclear bombs. Jihad operations in Yemen against communism will give jihad action in the Peninsula credibility and effectiveness" (AFGP-2002–600053). The far enemy, because he is far, is more likely to be threatened using international mass media. This is the preferred outlet for recordings regularly issued by Osama bin Laden, Ayman al Zawahiri, and Abu Musab al Zarqawi.

Tools of the Trade

Another sign of Islamist sophistication is their use of standard concepts, theories and best practices in their communication and media efforts. Perhaps the most impressive example of this is their use of after-action review. "A Memo to Sheikh Abu Abdullah" (AFGP-2002–003251), already quoted at length in this document, contains a detailed critical assessment of the failure of the jihadis to properly exploit their victory over the Americans in Somalia. "Lessons learned from the armed Jihad ordeal in Syria" (AFGP-2002–600080) attributes failure to win support for their efforts to lack of an acceptable public relations campaign. This and other documents we reviewed indicate that the Islamists do the functional equivalent of SWOT (strengths-weaknesses-opportunities-threats) analyses on their communication and media operations in an effort to improve quality.

There is also evidence of theoretical thinking about how effective communication campaigns are conducted. For example the "lessons learned" document just cited says:

> Obviously the most essential element of any revolutionary organization is putting forward a series of goals and slogans that attract the masses, and presenting itself as a revolutionary pioneering organization with crystal clear

> objectives. The true mujahideen failed to put forward their ideology, slogans and objectives via a well crafted media campaign.... such an explanation was and still is the main pillar for attracting the masses and mobilizing the base members on an intellectual and ideological level to partake in this dangerous work (i.e., Jihad). (AFGP-2002-600080).

This conclusion would not sound out of place in a modern social science classroom. For example it reflects a "accessibility-based" orientation to public opinion formation (Kim, Scheufele, Shanahan, 2002; Nisbet et al., 2004).

As for particular tools, in the 1990s al Qaeda ran a now-defunct paper called *Nashrat al Akhbar*. More recently we have seen the taped statements Islamists release on a regular basis. Videos of hostages and their executions can also be thought of as a kind of press release. In fact, a new group (or sub-group) has surfaced, calling itself the "Brigade of Media Jihad." Their mission is to intimidate the West through release of these videos, as well as videos of dead and wounded U.S. soldiers, via al Jazeera and the Internet (Trabelsi, 2005). The Islamists also place a great deal of emphasis on having good, succinct slogans. They were mentioned in the quote above, and also in "Five Letters to the Africa Corps" where al-Tajiki concludes that "Somalia needs a comprehensive national front that agrees on general, not detailed, Islamic slogans. That front must lead to an expanded national government approved by the major forces in the country: tribes and political groupings. The slogan acceptable to all is "Somali Freedom and Islamism" (AFGP-2002–600053).

Disinformation Operations

Islamist media strategy is also sophisticated in that it apparently makes use of disinformation. This mainly has to do with propagation of rumors that cast the apostates, Jews, and unbelievers in a bad light. Even though these reports fall on a scale somewhere between fantastic and absurd to the Western reader, the disinformation seems calibrated to appeal to the prejudices of the Islamists' audience, making them more receptive to legitimation and propagation arguments.

Disinformation serves the Islamists' legitimation strategy by casting the West as duplicitous and evil, as a force that Muslims are obligated to resist.

The Islamists and other terrorists have a long history of using this tactic. In the early 80s a rumor spread from Morocco to Indonesia that Neil Armstrong heard the call to prayer on the moon and converted to Islam but had to keep it secret for fear that he would lose his

U.S. government job (Mark Woodward, personal conversation). More recent examples are not quite as fantastic, but equally dubious. Rumors we have heard include:

- The U.S. Marine Corps barbequed Somali babies;
- 4000 Jews were warned by the CIA/Mosad not to come to work at the World Trade Center on 9/11; and most recently,
- The Indonesian tsunami was caused by a nuclear bomb detonated by the U.S. Navy.

These represent but a few examples that have gotten significant traction.

This disinformation serves the Islamists' legitimation strategy by casting the West (and the United States in particular) as duplicitous and evil, as a force that Muslims are obligated to resist. Disinformation has the advantage of being impossible to disprove. As a bonus, if the rumor catches on and spreads wide enough, it leads officials to issue defensive responses (as happened with the WTC/Jews rumor). These responses can then be claimed by the Islamists as even further evidence of Western duplicity.

Coordination with Operations

Islamist efforts are also sophisticated in that they are coordinated with operations. A specific example of coordination comes from one of the AQI documents, showing that the practice is alive and well among contemporary Islamists:

> There's how we prepare for him in few steps, like bringing him down through the media, then kill him, or kill his assistant who does the work for him, and see his reaction, then kill him, or fight him through the media, until his publicity is dead. (IZ-060316–01)

The thinking behind this coordination is explained by Abu Huthayfa:

> Political and informational functions are combined. Both are linked in the military function, and all which act together harmoniously to serve the ancestral jihad plan without dominance of one over the other. This is the obligation of the wise command, which administers and directs work with knowledge and charisma whereas it knows when to push politically, when to cool down the informational media, and when to kick off militarily. For each phase there is a plan and mechanism that fit its requirements and achieve the goals. (AFGP-2002–003251)

One of the mechanisms gaining popularity in the Islamist media strategy is the use of the Internet as a tool for disseminating information, as "the Web seems to be an appropriate technical infrastructure for communications and organizational design" (LaPorte, 1999, p. 215). Specific uses of this mechanism for both political and informational functions are described in more detail next.

Jihad and New Media

Our third point is that Islamists are technically savvy and intent on pushing jihad into the sphere of new media. By *new media* we mean electronic communication and information technologies other than traditional one-to-many broadcast methods of radio and television. While this could be considered another tool of the trade as discussed above, these media deserve singling out because they represent something of a convergence. On the one hand, al Qaeda has been transformed from a formal organization into a distributed social movement (Combating Terrorism Center, 2006a; Hoffman, 2004). Experts agree it is now more like a global network organization (Miles & Snow, 1986) than an army of fighters. The new media technologies that have exploded in the last decade are key enablers of global network organizations (Desanctis & Monge, 1999).

These facts are not lost on Islamists. Abu Huthayfa's letter describes several planned applications of new media. He goes on at some length about the need to develop a "huge database" that will support modern knowledge management practices that will enable the leadership to "manage struggle cleverly" and "carry on the appropriate strategies for each crisis." He also plans support for insiders when they need to collaborate and exchange information:

Islamists are technically savvy and intent on pushing jihad into the sphere of new media.

> Its [this cell's] main function is receiving the final product of the informational section and securely transmitting it in the country by an accurate Movement plan via using all available communication means. Personnel can be creative in utilizing secure methods of receiving and sending. For instance the electronic mail (e-mail) which may give a memory of up to (4MB) in sending attachment files along with the original message. This is a very fast method for sending the product of the informational section. Also, large web site locations to store files on the Internet are available for this purpose such as (www. Driveway.com), which ranges from (25MB) to (100MB). (AFGP-2002–003251)

The www.driveway.com reference illustrates the dangers noted above. The site is no longer in operation as such, and the domain name now leads to a site called remotepc.com, which is in turn owned by Pro Softnet Corp. of Woodland Hills, California. It is entirely plausible, even likely, that this company bought the domain name on the open market after its previous owner abandoned the registration. If so, then the owners and records of the site referenced in the quote are in all likelihood lost to posterity. This is quite suitable for the Islamists, as anonymous file sharing and so-called "warez" sites like the former driveway.com have mushroomed in recent years. They have no doubt moved on to a new favorite site, which will in time shut down and cover what few tracks they have left there.

Other applications of new media are aimed at outsiders. For example, the web is seen an essential mechanism for legitimation and propagation:

> The importance of establishing a web site for you on the internet in which you place all your legible, audible, and visible archives and news must be emphasized. It should not escape the mind of any one of you of the importance of this tool in communicating with people. (AFGP-2002–003251)

Still other Internet applications are aimed at intimidating the enemy, as the earlier example of the "Brigade of Media Jihad" illustrates. There are even more extreme examples of new media in Islamist practice: There are reports that Islamists have developed a CD of children's computer programs that mixes innocent games with firs-person shooter games featuring American targets (Brachman, in press).

The marriage of jihad with new media represents a dangerous development for several reasons. First, the Internet is, by design, decentralized and not subject to easy control by authorities. Second, laws have not yet caught up with the new media so there are many opportunities to operate outside regulation (as do offshore Internet gambling operations), and what regulations exist are inconsistent across countries. Third, the new media are (after all) new, so big government has not caught up to technology in many respects. In particular there are still many people in government—especially higher-up in government—who have little experience with new communication and information technologies and/or avoid using them for some reason. Fourth, any period of rapid development is chaotic, so new and unpredictable applications appear all the time. The result is an environment where Islamists can be anonymous, operate outside the law, exploit a shortcoming of their more ossi-

The marriage of jihad with new media represents a dangerous development.

fied opponents (an "asymmetry" in military parlance), and tap a stream of new developments in doing so.

The concern on the one hand is that Islamists will use new media as a tactical platform. Based on a netwars (Arquilla & Ronfeldt, 2001) scenario, decentralized but like-minded groups would use technology to coalesce, seemingly from nowhere, fight the authorities (who are not well structured to oppose such threats), and then dissipate just as fluidly. There is also concern on the other hand that new media will become a platform for "virtual jihad." Strong evidence exists that Islamists are moving into the new area of "social networking" services that have skyrocketed in popularity in recent years. "Most jihadist message boards on traditional websites are in Arabic and require users to know someone connected with the board before they can gain access. Social networking services such as Orkut, Friendster and MySpace, however, allow users to create personal profiles and associate with 'communities' based on shared interests" (Hunt, 2006).

Mediated communication is known for its ability to produce bonds between people who have never even seen one another. There is something of an epidemic in Japan of young adults meeting on the Internet and committing suicide together (McCurry, 2005). When non-fatal, relationships developed online can be even stronger than those developed face to face (Walther, 1996). Thus the new media have great potential as legitimation and propagation devices. As Brachman (in press) explains:

> One sees that the jihadi movement is not simply using technological tools to recruit new members, receive donations and plan attacks—all very real and serious threats. Rather, its membership is actually seeking to catalyze a computer-linked global social movement emerging from the very use of this technology; they are crafting a global counterculture based on the very process of participation as well as product. (p. 1).

Recommendations

Before giving recommendations based on our analysis, there is an important constraint to address. The credibility of the United States in the Muslim community (writ large) is perhaps at an all-time low. Statements from U.S. officials, such as President Bush calling the war on terror a "Crusade," have played directly into the hands of the Islamist communication strategy. They routinely frame the current U.S. presence in the Gulf region as a modern-day reproduction of the Crusades, call U.S. forces "the Crusaders," and so on. This taps negative collective memories (which in this part of the world seem longer and more robust than elsewhere) that help Islamists cast U.S. forces as untrustworthy invaders. Other incidents, such as prisoner abuse at Abu Ghraib and Guantanamo Bay, create rhetorical opportunities for Islamists to claim the United States is fundamentally no different from the defeated regime in Iraq and other apostate regimes in the region. Our inability to restore basic services in Iraq calls into question our capability to do what we say, whether or not our intentions are honorable.

The point of raising these cases is not to rehash criticism of U.S. actions. But regardless of whether these failures were preventable, they have seriously degraded U.S. credibility with Muslims in the Middle East and to varying extents worldwide (see Kohut, 2005). Given that our credibility with this audience was already low before the Global War on Terrorism, it is probably safe to say that it is currently in the range of slim to none. This places severe constraints on our ability to directly engage and compete with the strategic communication activities described above. For example, any messages designed to refute or undermine Islamist legitimation arguments are likely to be framed as further attempts to manipulate and corrupt Muslim minds.

This creates serious challenges for any counter-strategy designed to compete with the Islamists in the war of ideas, but it does not mean that we are helpless in the short term or that we should simply surrender the war of ideas to the Islamists. On the contrary, it is essential to resist their efforts, and there are indirect ways of doing so that do not depend on having high source credibility. We conclude this chapter by recommending a long-term effort to restore lost credibility, followed by a set of short-term strategies for competing with the Islamist communication and media strategy in the mean time.

1. Adopt a long-term strategy of improving our credibility with Muslim audiences.

Perceptions of credibility can be cultivated. According to existing research the main dimensions of credibility are trustworthiness, competence, and goodwill (see McCrosky and Young, 1981). If these dimensions generalize to Muslim culture, then we can improve credibility there rather straightforwardly by behaving in ways that the Muslim world views as trustworthy, that demonstrate our competence to accomplish outcomes that prove our goodwill.

This recommendation entails two challenges. First, it is a long-term effort that will not serve immediate goals of resisting Islamist efforts in the here and now. Restoring lost credibility requires an accumulation of experience over time that outweighs (or at least balances) the negative experiences of the recent past. This is an important constraint because until such balancing takes place there is little point in *overtly* competing with Islamist rhetoric or engaging in public diplomacy exercises. These efforts will simply not be taken seriously by the intended audience.

A second challenge is that our scientific and philosophical understanding of credibility is anchored firmly in Western culture, deriving from ideas of the ancient Greeks about effective participation in public discourse (e.g., Aristotle's *On Rhetoric*). It may be that exactly the same dimensions of credibility exist in Muslim ideas of public discourse, but to our knowledge no one has determined whether this is really the case. It is also possible that the same dimensions exist but are based on different evidence or experiences. In other words, what Westerners think indicates trustworthiness might not be the same as what Muslims think indicates trustworthiness. So our long term effort to restore credibility must include research into the generalizability of Western notions of credibility to the Muslim world.

2. Degrade Islamists' ability to execute their communication and media strategy.

The discussion above noted the Islamists' concern over having a simple, effective, coherent message. We also know from their after-action reviews that this is not an "automatic" occurrence. Previous efforts have, in the Islamists' own estimation, foundered on the failure to accomplish it. Having good PR is therefore a *challenge* for Islamist organizations. Doing it takes resources, coordination, and special effort. As with any organization, these things are in limited supply, meaning that events and priorities elsewhere in the organization can interfere or compete with the communication-related tasks. Therefore actions taken by the United States to create problems for Islamists on the organizational level should in turn impede their ability to formulate and deploy coherent messages.

Two recent publications offer suggestions about how we might pressure Islamist organizations. The "Harmony" report (Combating Terrorism Center, 2006a), which contained many of the Islamist texts analyzed for this chapter, is one. It concludes that the new "social movement" version of jihad creates significant problems of control for leaders, who encounter resistance from subordinates and allied groups. The difficulty of monitoring the actions of these deputies and allies creates conditions of suspicion and distrust that have the potential to degrade their organizational climate and ability for coordinated action. In another paper, Corman (2006) argues that all the activities of the Islamists are part of a structured, focused system. Some activities are compatible and complementary whereas others are incompatible or mutually exclusive. By viewing Islamist organizations as complex systems of these activities, it is possible to find combinations of inputs that cause maximum stress for the organization. Then operations can be undertaken to cause these kinds of inputs.

3. Identify and draw attention to Islamist actions that contradict Islam.

A key problem for Islamists is legitimating what they do, and as we said above the tenets of Islam provide rich sources of contradiction that complicate their legitimation efforts. An important way to compete with the Islamists then is to (a) identify these contradictions and then (b) make or encourage efforts to draw attention to them. One example, already mentioned, is the sin of killing Muslim children. The more we can promote knowledge of incidents where Islamists have killed children in the course of operations, the more the Islamists are challenged to "spin" these events in a way that maintains their legitimacy: "In the battle of perception management, where the enemy is clearly using the media to help manage perceptions of the general public, our job is not perception management but to counter the enemy's perception management" (Shanker & Schmitt, 2004, p. 1). Drawing attention to these discrepancies can also aid in implementing our second recommendation, as Islamist efforts toward perception management take resources away from other parts of the system.

Identifying the contradictions demands deep knowledge of Islam. It is necessary to identify passages in the Quran and associated religious texts that might call the Islamist actions into question. Since Islam is so decentralized (organizationally speaking) one must also know how these passages have been interpreted by relevant local/regional scholars. Because the Islamists are pursuing a legitimation strategy, their own efforts to drive the religious debate on certain issues might also tell us where they feel most vulnerable in this regard.

Once problematic actions and outcomes are identified, drawing attention to them requires (at minimum) a reliable source of information about when and where they happen. Returning to our example of child killings, reports of these incidents sometimes appear in news media, but we know of no source that systematically tracks the deaths in a detailed way. If there were, for instance, a web site that tracked children killed in Islamist attacks, making individual stories of the impacted families known, then relevant news media could be encouraged to make use of it. While the U.S. government would not be a credible host for such a site, it might be possible to encourage and support the development of one by a peace-oriented non-Salafi Muslim group.

4. Deconstruct Islamist concepts of history and audience.

Islamist ideology depends on a very particular construction of history. Their narrative begins during the golden age of the Caliphs and declines from there on. As Zerubavel (2003) notes, "historical plotlines are often extrapolated to imply *anticipated* trajectories" (p. 17). This means that if an audience accepts such a narrative, then they are apt to believe that as bad as things are now, they can only get worse in the future. This promotes the belief that going back to the past will solve all problems and provides a built-in logic for rejecting anything in the present associated with the decline. Orienting to this idealized past simultaneously helps solidify identity, creates a sense of legitimacy, and sets an unambiguous path ahead.

It is possible to generate counter-narratives to this one of decline. As Zerubavel explains, "we are not dealing here with actual historical trends but with purely mental historical outlooks. The very same historical period, after all, is remembered quite differently depending on whether we use a progress or a decline narrative to recount it" (p. 16). More research and analysis would be needed in order to shape a counter-narrative of progress. However introducing it into the public discourse would call into question the wisdom of returning to the past that the Islamists value so much, helping to undermine the very foundation of their ideology and disrupt their legitimation efforts.

In a similar way, Islamist communication and media strategy depend on a particular construction of the audience, described above. Blurring the lines in this scheme makes messages designed for particular audiences less effective. In the audience concept shown above, the sympathizers and troublemakers are the ones at risk because the bad guys are already condemned. If troublemakers can be convinced that they will sooner or later be viewed as the bad guys, and sympathizers can be convinced that they could easily be reclassified as troublemakers, then the good guy end of the audience concept begins to look less like safe categories of tolerance and more like a process that transforms good guys into bad guys.

5. Redouble efforts to engage Islamist new media campaigns.

Our review shows that Islamists are increasingly moving their communication operations out of the mass media and into the new media, especially the

Internet. This provides them with the ability to coordinate actions and build community even if they are not concentrated in geographical space. Their abilities in this area represent an asymmetry, as many Americans who did not grow up with the Internet still do not understand it well. This includes analysts, policymakers, officers, and so on who are responsible for fighting the Islamists. Without background knowledge of Internet "culture" it is difficult to understand how it might be used or keep up with innovations that Islamists can adapt for their purposes.

Leveling this playing field demands creative action. We envision creation of a permanent "geek battalion" dedicated to understanding, monitoring, disrupting, and counteracting Islamist Internet activities. This unit would include to as great an extent as possible young people recruited for their knowledge of Internet culture and technology. There are informal "roles" in this culture that could be particularly useful for this effort. For example "white hats" are hackers who use their skills for good, to catch other hackers, reveal security weaknesses of systems to their operators, and so on. Certain people are discussion board and chat room mavens who are good at knowing who is talking about what in which locations. Internet porn and warez traders would naturally know all the latest tricks and resources for exchanging files anonymously. A team composed of such people, especially if combined with relevant language skills, would go a long way toward reducing the asymmetries the Islamists now enjoy.

We also note that while the Internet may serve the Islamists they are also vulnerable to its foibles. For example, Islamist web sites are purposely of obscure origin, so there is no good way to distinguish a real one from a fake one. Fake web sites might be used to claim responsibility for deaths that Islamists would rather see blamed on others, as in the AQI example cited above. It is also possible through various techniques (well known to spammers) to "hijack" visitors to various web sites and take them to a different one with a competing message. Given operators with the proper knowledge, an array of such tactics could be deployed to make Islamist Internet use less effective.

6. Make better use of sympathetic members of the American Muslim community.

In closing we point out that an untapped resource in this effort is the American Muslim community. With the possible exception of recommendation 2, they are in a better position than almost anyone else to understand how to execute these strategies. Help from this community would be key in developing a "Muslimized" theory of credibility. They are better equipped than any analyst or non-Muslim academic to identify the tenets of Islam that are most at odds with Islamist strategy and actions and understand which of these contradictions are most likely to resonate with Muslims in the Middle East and elsewhere. They would be essential in framing a narrative of progress to undermine the Islamist narrative of decline. And an effective geek battalion would have to be rich in Arabic speakers who understand Muslim culture. In our view it is past time to start making use of this untapped resource in the war of ideas.

Conclusion

This chapter shows that Islamists place a great deal of emphasis on developing comprehensive media and communication strategies to aid their side in the war of ideas. However, the Islamists' ability to implement such strategies is not well understood and has been "systematically undervalued." Sophisticated media strategies aimed at themes of legitimating, propagating and intimidating serve the Islamist short- and long-term interests of driving invaders from the Arabian Peninsula and restoring and expanding an Islamic Caliphate. Their strategies are crafted after careful audience analysis and message adaptation, two of the most fundamental rules underlying any communication or public relations campaign. The increasing use of new media, such as the Internet, aid the Islamist cause further, by allowing an asymmetrical operating environment from which information may be disseminated while maintaining organizational security. While these practices and strategies are strong, it does not mean that they are infallible or irresistible. The recommendations we give in this chapter revolve around the ideas of establishing credibility and exploiting weaknesses and contradictions in Islamist messages. The resources to be a contender in this media war are within our reach, but we need to mobilize these resources and use them to our advantage.

Note

1. In the earlier white paper version of this chapter, in the title and throughout, we used the term *jihadi* to describe the terrorist forces. Since that time we have been convinced by arguments that use of this term constructs the terrorists as holy warriors, facilitating their religious legitimation strategy. Therefore, we have replaced the term *jihadi* with the *Islamist*, except in quoted references. The new term im-

plies an ideology rooted in a particular religious viewpoint, without implying that the religion itself is corrupt (any more than the term *sexist* implies that *sex* is bad). It does this without legitimizing the viewpoint in question.

References

Arquilla, J., & Ronfeldt, D. (2001). The advent of netwar (revisited). In J. Arquilla & D. Ronfeldt (Eds.), *Networks and netwars: The future of terror, crime, and militancy* (pp. 1–25). Santa Monica, CA: Rand.

Bin Laden Tape (2006, January 20). Text: Bin Laden Tape. BBC. Available online: *http://news.bbc.co.uk/2/hi/middle_east/4628932.stm*

Brachman, J. (in press). Al-Qaeda: Launching a global Islamic revolution. *The Fletcher Forum of World Affairs.*

Combating Terrorism Center (2006a). *Harmony and disharmony: Exploiting al-Qa'ida's organizational vulnerabilities.* Report by the Combating Terrorism Center, United States Military Academy. Feb. 2006. Available online: *http://www.ctc.usma.edu/aq.asp*

Combating Terrorism Center (2006b). *Al-Qa'ida's in Iraq hampered by bureaucracy and loss of Sunni support.* Report by the Combating Terrorism Center, United States Military Academy. April 2006. Available online: *http://ctc.usma.edu/CTC%20—%20Zarqawi%20Letters%20Analysis.pdf*

Corman, S. R. (2006). Using activity focus networks to disrupt terrorist organizations. *Computational and Mathematical Organizational Theory, 12(1)*, 35–49.

Desanctis, G. & Monge, P. (1999). Introduction to the special issue: Communication processes for virtual organizations. *Organization Science,10(6)*, 693–703.

Hoffman, B. (2004). The changing face of Al Qaeda and the global war on terrorism. *Studies in Conflict & Terrorism, 27*, 549–560.

Hunt, K. (2006, March 8). Osama bin Laden fan clubs build online communities. *USA Today.* Available online: http://www.usatoday.com/printedition/news/20060309/a_google09.art.htm

International Crisis Group (2006, February 15). *In their own words: Reading the Iraqi insurgency.* Middle East Report N°50. Available online: *http://www.crisisgroup.org/home/index.cfm?id=3953&l=1*

Kim, S., Scheufele, D. A. and Shanahan, J. E. (2002). Agenda-setting, priming, framing and second-levels in local politics. *Journalism and Mass Communication Quarterly, 79(1)*, 7–25.

Kohut, A. (2005, November 11). *How the United States is perceived in the Arab and Muslim worlds.* Testimony before the U.S. House International Relations Committee, Subcommittee on Oversight and Investigations. Available online: *http://pewglobal.org/commentary/display.php?AnalysisID=1001*

LaPorte, T. (1999). Contingencies and communications in cyberspace: The world wide web and non-hierarchical co-ordination [Electronic version]. *Journal of Contingencies & Crisis Management, 7*(4), 215–225.

McCurry, J. (2005, March 2). Seven die in online suicide pact in Japan. *The Guardian*. Available online: http://www.guardian.co.uk/japan/story/0,7369,1428256,00.html

McCrosky, J., & Young, T. (1981). Ethos and credibility: The construct and its measurements after two decades. *Central States Speech Journal, 32,* 24–34.

Miles, R. E., & Snow, C. C. (1986). Organizations: New concepts for new forms. *California Management Review, 28*, 62–73.

Nisbet, et al. (2004). Public diplomacy, television news, and Muslim opinion. *Press/Politics, 9*(2), 11–37.

Rumsfeld, D. (2006). *New realities in the media age: A conversation with Donald Rumsfeld.* Council on Foreign Relations. Available online: *http://www.cfr.org/publication/9900/*

Shanker, T., & Schmitt, E. (2004, December 13). Pentagon weighs use of deception in a broad arena [Electronic version]. *New York Times.*

Trabelsi, H. (2005, August 19). Al-Qaeda group to terrorise US. *News24.com* (AFP wire). Available online: *http://www.news24.com/News24/World/News/0,,2–10–462_1756675,00.html*

Walther, J. B. (1996). Computer-mediated communication: Impersonal, interpersonal, and hyperpersonal interaction. *Communication Research, 23*, 3–43.

Zerubavel, E. (2003). *Time maps: Collective memory and the social shape of the past.* Chicago: University of Chicago Press.

CHAPTER FIVE

The Iranian Letter to President Bush

Analysis and Recommendations

H. L. GOODALL, JR., LINELL CADY, STEVEN R. CORMAN, KELLY MCDONALD, MARK WOODWARD, AND CAROLYN FORBES

Introduction

On May 9, 2006, world media outlets released news of a letter written by Iranian President Mahmoud Ahmadinejad to U.S. President George W. Bush. The letter was the first official communiqué from the Iranian government to the U.S. since the two countries broke diplomatic ties in 1979. The letter was dismissed by U.S. spokespersons as a "rambling" narrative or as a "meandering screed" that did not address the current U.S. concerns over the nuclear energy program initiated by President Ahmadinejad. For the next few days, world media sources repeated the U.S. dismissal while offering their own assessments of the meaning and significance of the communiqué.

Reactions to the letter were mixed. Controversies over how the incident was handled pointed to the "unsophisticated response" made by U.S. officials to the overture, however diplomatically unorthodox its format. Sources throughout the world indicated that not only had the intention of the letter been misrepresented by U.S. officials, but its meaning had also been misinterpreted, thus fueling ongoing speculation that any issues raised by President Ahmadinejad were secondary to the stated U.S. goal of discussing only nuclear development in Iran.

We answer three key questions: (1) How did the controversy play in international media outlets? (2) What was the intention and content of the letter, and was it, in fact, a "meandering screed?" (3) What lessons may be derived from this incident to guide future decisions about U.S. strategic communication?

Our analysis provides the following conclusions:

- The letter is addressed to President Bush but is intended to reach a broad international audience—perhaps all believers. As such, it represents a seemingly distinctive ecumenical approach to organizing all monotheist religions against the evil influences of Western style democracy and liberalism.
- The letter is not a "meandering screed" but instead is an organized and coherent statement that provides a focused narrative.
- The letter represents a *dakwah* or invitation to President Bush, which may be interpreted as a call to Islam and/or a prelude to violence.

Two broad recommendations are generated from our analysis: (1) the U.S. needs to develop a more theoretically and culturally informed *independent* process for analyzing and managing diplomatic communication; and (2) the U.S. needs to open communication with President Ahmadinejad by formulating a response to the letter in order to improve our image with other Muslim audiences around the world.

THE LETTER AND ITS RECEPTION

Release of the Letter

On May 9, 2006 media outlets worldwide published news stories about the existence of a personal letter written to President George W. Bush by Iranian President Mahmoud Ahmadinejad. The letter, 16 pages in Farsi and accompanied by an 18-page English translation, addressed what "an Iranian spokesman called 'new ways' to resolve the crisis over Iran's nuclear program."[1] The letter had been delivered to President Bush the previous weekend. It was also reported widely that the letter was the first official communication from the

Iranian government to the United States since diplomatic ties were broken between the two countries in 1979.

United States spokespersons were quick to dismiss the relevance of the letter to current concerns about Iran's plans to enhance its nuclear program:

"It was a meandering screed."

Unnamed U.S. official

> "This letter isn't it," Secretary of States Condoleezza Rice said in an interview with The Associated Press in New York. "This letter is not the place that one would find an opening to engage on the nuclear issue or anything of the sort. It isn't addressing the issues that we're dealing with in a concrete way."

An unnamed U.S. spokesperson was widely quoted saying "it was a meandering screed." John Bolton, U.S. Ambassador to the United Nations, added that with this letter "Iran was throwing 'sand in the eyes' of diplomats."

By May 10, 2006 a full-text translation provided by a French source (*le Monde*) found its way into public circulation (see Appendix A). Media outlets from around the world reported the existence of the letter as well as the repeated dismissal of its relevance to the current debate about Iran's nuclear program by U.S. officials. On that day during an interview in Florida with representatives from seven newspapers, President Bush was asked about the letter and he replied: "It looks like it did not answer the main question that the world is asking, and that is, 'When will you get rid of your nuclear program?'"

Also on May 10 *The New York Times* published a story about the letter that included the following account:

> "Mr. Ahmadinejad reiterated that Iran would reject any Security Council decision that restricted the country's nuclear activities, which he said his country, like others, had the right to pursue.
>
> "'The Iranian nation has decided,' he said at a news briefing in Indonesia that was carried by the Iranian news agency IRNA. 'It will defend and never renounce its rights.' Mr. Ahmadinejad also cast the tensions over Iran's nuclear program as an unfair struggle against technological advancement in Muslim countries.
>
> 'Iranians are strong enough to defend their rights,' he said. 'But it should also be stressed here that resistance of the Iranian nation will not only be for Iran but for all independent-minded states including Indonesia, Malaysia, Turkey, Egypt and other Muslim countries.'"

Foreign Media Coverage

The reaction by the foreign media to the letter to President Bush from President Ahmadinejad was swift and voluminous in nature.[2] It is striking that so many of the sources comment that it is irresponsible to simply dismiss the communication, as it represents a rare diplomatic opening in an otherwise silent quarter of a century. The letter, argued many sources, presents a unique opportunity for dialogue between the two nations and their leaders.

Though doubtless alarmed by some of President Ahmadinejad pronouncements about Israel, one writer for *The Jerusalem Herald* concluded, "**it would** be wrong to dismiss Ahmadinejad's letter to Bush as just another of the Islamic leader's many weird habits. It would be more prudent, and better politics, to take Ahmadinejad seriously and to try and understand him on his own terms" (*emphasis in original*, Taheri, 2006). Foreign commentators view the stakes for both sides in the escalating political war of words as gravely serious. Dr. Marwan Al Kabalan, a lecturer in Media and International Relations on the faculty of Political Science and Media, Damascus University, said the letter is significant in that "the Iranian leadership must have realised (sic) how dangerous this game is and hence may have decided to approach the Americans and get them to talk about what concerns them" (Al Kabalan). *The Daily Star* of Lebanon noted, "For the first time in years, there is cause for hope. . . . The most effective way to resolve the international standoff over Iran's nuclear program and one of the few remaining strategies that has not yet been tried is through direct talks between Tehran and Washington" (*A.Q.I.*, Kole).

> **"It would be wrong to dismiss Ahmadinejad's letter"**
>
> **The Jerusalem Herald**

These reactions stand in sharp relief to coverage of the U.S. reaction to the letter. Media outlets widely covered the dismissive remarks by administration officials and the cool reception by the White House. Reporting on Secretary of State Rice's remarks on weekend television about the letter, *Agence France Presse* led with Secretary Rice's intonation that the letter "[I]s not a serious diplomatic overture" (AFP, *LN*). From a State Department press briefing, *Xinhua News Service* highlighted the remarks by spokesman Sean McCormack who noted, "Our view at this point is that there are plenty of channels of communication if the Iranians want to pass information to us or we want to pass information to them" (*Xinhua, LN*). Still, other sources, noted the Washington official's remarks that the letter did little to thaw the chill in relations between the U.S. and Iran as the Administration's questions surrounding Tehran's nuclear intentions remained unanswered (Cornwell, 2006).

While political and policy differences within Iran were highlighted by some (Al-Issawi; *Mideast Mirror*), the logic and position of Iran—with regard to the United States—was highlighted by more sources, identifying comments by both President Ahmadinejad and those in his government (*BBC Monitoring(a); Deutsche Presse-Agentur; Emirates News Agency*). Further, the letter's open invitation for conversation, as well as chastisement, was highlighted by some of the same sources, noting areas of shared national and international urgency (*Mideast Mirror; BBC Monitoring (b)*).

Analysis of the Debate

Within a span of a few short days, the first communiqué from Iran to the United States in a generation had been framed solely in relation to the current nuclear development issue. Reaction to the letter by foreign media was highly favorable to the position of the leadership in Teheran and tended to frame the reaction by American officials as dismissive and somewhat unsophisticated given the detailed and extended prose of the letter. Drawing the parallel to the letter written by the late Ayatollah Ruhollah Khomeini to then Soviet Premier Gorbachev, one writer remarked that the letter needs to be read and understood in its proper context as both a political and religious document (*La Guardia*).

Whether or not this letter was prompted by the escalating debate over Iranian nuclear weapons development, it seems clear that its scope is much broader. While the letter is extremely critical of the U.S. and accordingly provides ample grounds for criticizing Iran and President Ahmadinejad, there are also many portions that highlight possible areas for rapprochement, even cooperation, in issues of mutual concern. On this singular occasion of a personal letter from the President of Iran to the President of the United States, were our official responses to it both accurate and appropriate? Was the letter really a "meandering screed?" What are the diplomatic implications of such dismissals given the broader understanding we have acquired about communication emanating from Middle Eastern religious and political leaders?

The challenge, for both the Presidents involved and everyone else, is to read through confrontational rhetoric to discover the possibilities for dialog. Our purpose in this chapter is to examine the Iranian letter within the context of current understandings of intercultural communication and diplomacy. First, we will provide an analysis of the *intention and context* of the letter and our official responses to it within the broader interrelated domains of world media circulation, world attitudes toward the U.S. and the Bush administration, and religious thought. Second, we will offer an *analysis of the content of the letter* to see if, in fact, it can fairly be characterized as a "meandering screed" and if the interpretations of its meanings have been accurate, given what is known about Middle Eastern cultures and the Muslim religion. Third, we will offer *strategy and process recommendations about future communication opportunities with Middle Eastern leaders*. Our goal is to help our leaders and spokespersons apply the wealth of current thinking about communication, culture, and religion to practical problems affecting the image, honor, and standing of the U.S. in the world.

Intention and Context

President Ahmadinejad frames the letter to President Bush with this question:

> My basic question is this: Is there no better way to interact with the rest of the world? Today there are hundreds of millions of Christians, hundreds of millions of Muslims, and millions of people who follow the teachings of Moses. All divine religions share and respect one word, and that is monotheism, or belief in a single God and no other in the world.

While questions of intention are always tricky, and made more so by issues of cultural and language differences, nevertheless the above paragraph provides a clear identification of a major theme that recurs in the letter and frames the issues informing it. Moreover, the statement provides two important clues to the meaning of the letter: (1) interaction with the religious peoples of the world is flawed and must be addressed if relations among nations and tribes are likely to improve; and (2) all religions share a belief in a single God—or, put differently—there is a unifying power available to those who want to interact more effectively with religious people in the world.

Key theme: Interaction with the religious peoples of the world is flawed and must be addressed.

President Ahmadinejad marshals an array of cultural, political, economic, and religious arguments to support his contention that "interaction" with religious peoples of the world is in danger and must be repaired. Those arguments include themes that are neither new nor particularly distinctive from those articulated by Osama bin Laden or other Muslim religious leaders and politicians since 9/11. His arguments, however, do include new nuances and ecumenical references that inform and deepen our analysis.

President Ahmadinejad's major themes are:

- There are contradictions between a professed belief in God and the word of his prophets and the current political, economic, social, and religious behavior of the Western democracies, particularly the U.S. The U.S. operates prisons at Guantanamo and elsewhere without providing inmates with legal representation or trials while keeping them away from their families;
- The continued existence of Israel and the unfailing support of the U.S. for that nation poses problems for other nations in the Middle East. Not only do we not support the democratically elected officials of Palestine, but the U.S. government's use of the existence of Israel to call technological advances by others in the region a threat to their existence;
- The U.S. government tells lies to its citizens and to the rest of the world;
- Other nations in Latin America, Asia, and Africa should be of greater concern to the U.S., and we should be helping them eradicate disease, end poverty, and provide peace and security to their citizens;
- The U.S. has used the "horrendous incident" of September 11, 2001 to perpetrate a mediated culture of fear and anxiety, to avoid telling the truth, and to turn away from the teachings of God. The U.S. is spending "hundreds of billions of dollars" on war when even within the U.S. poverty, homelessness, and unemployment exist.
- Is George W. Bush not concerned how he will be remembered by the people he has led? Did he manage to bring peace, security, and prosperity to his people?
- While "it is not my (President Ahmadinejad's) intention to distress anyone," if the prophets of all the world's great religions were with us today, how would they judge such behavior?
- "Liberalism and Western style democracy . . . have not been able to realize the ideals of humanity. Today, these two concepts have failed." There is a worldwide religious movement toward God. Will (President

Bush) not join them?

Toward the end of the letter, President Ahmadinejad issues an "invitation" to President Bush:

> Will you not accept this invitation? That is, a genuine return to the teachings of prophets, to monotheism and justice, to preserve human dignity and obedience to the Almighty and His prophets?

As we will discuss later, this specific "invitation" (i.e., *dakwah*) has a known religious and political context within the Muslim world. So, too, do the concepts of "justice," "human dignity," and "obedience to the Almighty and His prophets."

The letter reflects political themes being aligned across the Middle East, Europe, and Asia.

On the surface the letter has little to do with the chief U.S. agenda item—Iran's nuclear capability. It has everything to do with the major political themes increasingly being aligned across the Middle East, Europe, and Asia by political, religious, and jihadi leaders. President Ahmadinejad is merely the most recent spokesperson. However, that he chose to break diplomatic silence with a personal letter to an American President and reiterate these themes deserves much closer scrutiny. Is it reasonable to interpret the letter as the ramblings of a dangerous and perhaps deranged political foe who is trying to deflect attention from the nuclear issue? Or was it—indeed is it—a window of diplomatic and cultural opportunity? Our analysis focused on two related questions: (1) is it reasonable to interpret the *form* of the letter as a rambling screed, and (2) was the *content* of the letter fully and accurately interpreted?

Form: Was it a "Meandering Screed"?

As we have documented an unnamed "White House official" said that Ahmadinejad's letter was a "meandering screed," a characterization that was picked up in the wider press and sometimes transformed. For example, A UPI release (May 9) called it a "rambling letter." From a rhetorical point of view, characterizing the text as "meandering" or "rambling" is highly significant because it has implications for how seriously this message should be taken.

The letter is 3% more focused than the average news story even though it is much longer.

While each reader can decide for himself or herself whether the text is a rambling screed, in most cases this is a politically loaded judgment. We therefore turned to a more objective method of formal text analysis to determine how reasonable it is to describe the letter as "meandering" and "rambling." Such characterizations essentially claim that a text is unorganized. Unorganized texts drift from topic to topic without taking care to connect the ideas being discussed into a coherent framework. A text analysis technique called Centering Resonance Analysis (Corman, Kuhn, McPhee, & Dooley, 2002) allows us to measure how organized Ahmadinejad's letter is in a standardized way that permits comparison to other examples of communication. The measure, called *focus,* gives a value between zero and one describing the extent to which the concepts in a text are systematically organized. An important point about focus is that, other things being equal, longer texts tend to be less focused because they use more concepts, and it takes more effort/skill to organize more concepts.

The focus score for Ahmadinejad's letter is 0.387. For comparison, the average focus of news story published by Reuters news service in the first quarter of 2006 was 0.358. So the letter was about 3% more focused than a news service story. This is significant because the letter was considerably longer than the average news story.

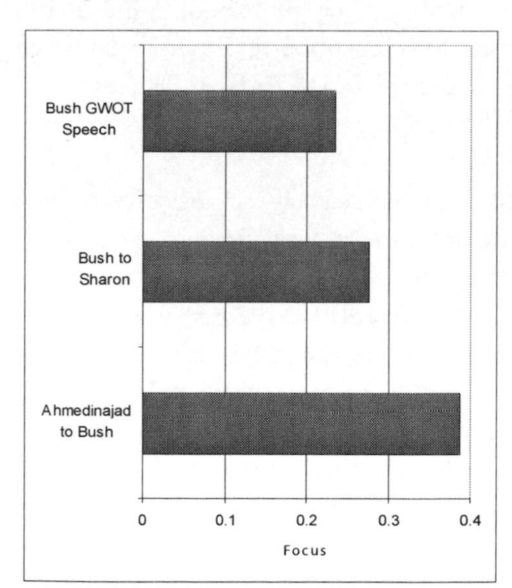

Figure 1. Comparative focus of three speeches.

As another point of comparison we consider two recent texts from President Bush. One example is the text of a letter from Bush to Israeli Prime Minister Arial Sharon in April 2004 dealing with a possible settlement between Israel and the Palestinians. The second is the text of a speech on the Global War on Terror delivered April 6, 2006 delivered in North Carolina. The letter, though shorter than Ahmadinejad's letter to Bush, is a good point of comparison because it is a similar genre of communication. The speech is comparable in length to Ahmadinejad's letter and is on a related topic.

Figure 1 shows the comparative focus of these three texts. Ahmadinejad's letter is the most focused of the three texts. It is over 10% more focused than

Bush's letter to Sharon, despite the fact that is it almost twice as long. It is 15% more focused the Bush's GWOT speech, which is of comparable length. Clearly nobody would characterize these texts of President Bush as "rambling" or "meandering." By the same token, it was unreasonable for the "White House official" to characterize Ahmadinejad's letter as such given that it is considerably more focused.

Content: Was the Letter Fully and Accurately Interpreted?

As noted above, this letter is an elegant expression of *dakwah*, a call to Islam. It underscores the suffering and oppression that afflict many of the world's populations in the current global order, and hopefully points to the transformative power of divine justice that may result from more faithful obedience to God. There is also an apocalyptic undertext about the second coming of Christ. Ahmadinejad's apocalyptic sympathies have been pointed out before, as have those of some of bin Laden's supporters. It is important to note, however, that this is consistent with an orthodox interpretation of Islam. Many in the Middle East believe the situation in the region is so dire that this is the only hope for a solution.

The letter is an elegant expression of *dakwah*, a call to Islam.

Although the letter is addressed to President Bush, it is clearly directed at a larger audience. Perhaps most fundamentally, the letter is directed at the Muslim world as a whole. It is significant that there are no Shiah sectarian themes in the document. For example, there are no references to Imams, central figures in Shiah Islam but not in Sunni Islam. Even more significantly, perhaps, the apocalyptic portions of the text do not reference the return of the "hidden" Imam who figures significantly in Shiah apocalyptic thought.

Dakwah is traditionally directed to Muslims, and this letter may reflect Ahmadinejad's invitation to Bush to become Muslim. (That the invitation to become a Muslim is traditionally viewed as obligatory prior to the use of violence has been noted by some interpreters.) He uses what would appear to be classical Islamic religious/political vocabulary, including peace, justice, tranquility, responsibility, dignity (see Bernard Lewis *The Political Language of Islam*) and defines the world's problems as the result of leaders who disobey God. As he rhetorically asks: "My basic question is this: Is there no better way to interact with the rest of the world?"

It is also noteworthy that the letter emphasizes the shared Jewish, Christian and Muslim values. In his words: "Today there are hundreds of millions of Christians, hundreds of millions of Muslims and millions of people who follow the teachings of Moses (PBUH). All divine religions share and respect one word and that is monotheism or belief in a single God and no other in the world." The letter is anti-Israel but not anti-Jewish. He takes the classical Islamic view that Jews, like Christians are "people of the book" to whom God has sent Prophets and Holy Books. The Jewish Prophet to whom he refers is Moses (Musa) and the book is the Torah. Muslims also consider Jesus to be a Prophet and the Gospels to be the revealed book of the Christians.

The more ecumenical dimensions of this version of *dakwah* suggest that Ahmadinejad may be seeking an even larger audience. The classical expression "Peace Be upon Him" after each reference to the prophets, including Jesus and Moses, exhibits a respectful tone. One most commonly hears this phrase after the name of Mohammed. So the many references to Jesus, followed by this phrase, in one single text are rhetorically very powerful. We are not aware of another Muslim text with so many references to Jesus with this ritual address of honor in such a condensed space. Consequently, the invitation to Islam is modified, seeming to extend this classical genre in a broader way to include the three religions of the Book and indeed the God of all people.

To this extent, the letter is an interesting intervention in the "clash of civilizations" dynamic. The "Islam against the West" framing that has been advanced by many, including especially the jihadis, shifts slightly insofar as the dividing line is between faithful followers of God (not simply followers of Islam) and the godless. This is only a slight variation on the former narrative, however, insofar as the primary political forms within the West—liberalism and Western style democracy—are denounced as failed concepts. He argues throughout the letter that secular democracy leads to war, oppression, misery, and poverty. He insists that the shared values of Islam, Judaism, and Christianity hold the key to a transformed world order that is more peaceful, just, prosperous, and dignified. What counts as faithful obedience to God remains captive to Ahmadinejad's conservative version of Islam, a version that he proclaims is the shared values of Islam, Judaism, and Christianity. But the ecumenical manner in which he makes this case may well make it more compelling to larger segments of the Muslim world who reject jihadi versions of "Islam against the West" yet still feel alienated.

The ecumenical nature of the letter may make it more compelling to non-jihadi Muslims.

Dismissing this letter as the ramblings of a religious fanatic fails to honor the cultural resonance and legitimacy of this genre within the Muslim world. Not only may such a response intensify the strong anti-American feelings among Muslims, it may constitute a major "missed opportunity" to state our case in the public forum. In addition to acknowledging our shared values that are rooted in our religious heritages, it is essential to challenge the alignment between faithful obedience to God and rejection of liberalism and democracy. The rhetorical power of the letter rests upon setting these in opposition, and this must be directly countered.

The "wall of separation" between church and state in the American tradition sets constraints on the response that the United States can realistically make to this letter. But that does not mean that it need avoid a substantive response. An effective response would make some reference to our shared values and express sympathy for the suffering and oppression that mark the lives of so much of the world's population. It would acknowledge this reality of the current global order, but it would then move on to disrupt the letter's alliance between liberalism, democracy, and godlessness.

The privatization of religion within western democratic societies is sometimes accompanied by a sense that religious rhetoric carries no public claim, that it is essentially "irrational" and a form of irritating moralizing. The globalized world we are coming to inhabit makes it increasingly important to recognize that there are quite different rules regarding the appropriate role of religion in public discourse in other countries and traditions. After all the U.S. has a long rhetorical tradition of mixing prophetic religion with the mission of liberalism and democracy. Ahmadinejad's letter brings this issue to center stage.

Summary

Our analysis of the communicative, cultural, political, and religious aspects of the letter provides the following conclusions:

- The letter is not a "meandering screed" but instead is an organized and coherent statement that provides a focused narrative, which is at least as organized as statements typically made by President Bush;
- The letter represents a *dakwah* or invitation to President Bush, which may be interpreted as a call to Islam and/or a prelude to violence;
- The letter is addressed to a broader international audience—perhaps all believers—and as such represents a seemingly distinctive ecumeni-

cal approach to organizing all monotheist religions against the evil influences of Western style democracy and liberalism.

Recommendations: Process, Strategy, and Plan

From a communication perspective, the letter represents an interesting rhetorical intervention in a situation defined by mediated conflict, disagreement over principles and methods, cultural misunderstandings, and open hostility. This letter—all 18 pages of it, where not a single sentence refers to the development of nuclear technology—provides President Bush with a unique opportunity to engage President Ahmadinejad (and by extension, Muslims worldwide) in a personal as well as national conversation about how people of different faith traditions, different political systems, and differing views even on Israel can learn to live in peace, prosperity, justice, and mutual respect. The letter, and the official U.S. response to it, should serve as a case study in failed strategic communication. By misreading the intent and ignoring most of the content of the message, by failing to adequately interpret the meaning of the letter within known intercultural and religious frameworks, and by refusing to respond to the significance of the communiqué even after world opinion chimed in, the U.S. lost an important opening for dialogue and further tarnished our image on the globalized world's mediated stage.

Lessons About Process

One consistent challenge faced by official U.S. spokespersons is the need for an informed approach to a message analysis that is strategically coupled with a coordinated approach to message alignment across government, military, and contracted spokespersons and agents. The Iranian letter incident provides a perfect example of the problems that occur when neither informed analysis nor alignment of messages occurs but also affords those who are committed to improving communication of and about the U.S. an instructive heuristic to guide future actions.

The first lesson is the clear and present need for a theoretically and culturally informed non-governmental agency to provide real-time intercultural communication and religious analyses of messages *prior to* the formation of a coordinated strategic response. Relying on existing government agencies to provide communication interpretation and message strategy is a flawed policy because it means tying analysis to extant political agendas and therefore risks the imposition of political power dynamics to the interpretation of intercultural messages.

The second lesson is that the U.S. should frame these kinds of communiqués as opportunities for *dialogue*. From the religious philosopher Martin Buber forward, the idea of dialogue has been used to underscore "a meeting of the minds" mediated by "open, honest, authentic communication" that requires both participants to be "profoundly open to change" (Eisenberg, Goodall, & Trethewey, 2006).

By providing an 18-page letter organized by key questions articulated around common Middle Eastern concerns, President Ahmadinejad opened a line of moral and philosophical communication that appears to be aimed at cross-cultural dialogue with President Bush on vital issues of the day.[3] To ignore it, or to dismiss it, is to suggest that the invitation is either unacceptable or unworthy of respect. Either interpretation is unlikely to win the hearts and minds of Muslims in the Middle East (and elsewhere) who understand the cultural, political, and religious dimensions of this message.

Lessons About Implementing a Strategic Response

However President Bush chooses to respond to this invitation to talk, we believe the following principles should guide the resulting interaction:

1. Communicate directly to President Ahmadinejad using the medium he selected—an open letter. Deliver the letter in English and with a careful Farsi translation.
2. Demonstrate respect for the courage to initiate direct contact after such a long hiatus. Point out that the letter raises deep and important questions and we value an open exchange of ideas.
3. Invite President Ahmadinejad to a summit at Camp David to further engage these ideas.

The above recommendations will accomplish two important strategic goals: (1) it will provide clear evidence of a willingness to engage world leaders on questions they have about our intentions in the Middle East, (2) it will place the "talking stick" back in the hands of President Ahmadinejad, and (3) it will signal to Muslim audiences around the globe that the U.S. President is willing to engage their leaders on matters of mutual interest beyond the singular nuclear development controversy. An added bonus will be a clear signal heard around the world that the United States understands the importance of cultural understanding and human communication—of a personal as well as the mediated variety—in winning the global war on terror.

Robert Wright, in a review of two new books about anti-Americanism in the world, says this: "So history has put America in a position where its national security depends on its further moral growth." Our task, in part, is "to learn to see ourselves as others see us," which is to say, to understand how the have-nots of the world see the haves. Throughout the Middle East, moral leaders are often religious leaders, and religious leaders often rise to lead their nations. In our country, we must look to our President to provide moral leadership as well as to guard the interests of the United States in world opinion. To do that will require a renewed dedication to a very American principle: *What we say matters.*

We believe our President may be at a tipping point in his own appreciation of the power of words to change perceptions of reality and to heavily influence tensions in the Muslim world. In a recent press conference in Berlin, he said: "We must understand words mean different things to different people, . . . Sometimes my own messages send signals that I don't mean to send, but stir up anxieties in the Muslim world" (White House, 2006). Our hope is that with a more strategic understanding and use of communication, this administration and its spokespersons may halt the continuing deterioration of our image worldwide. Responding to the Iranian letter would be an excellent step.

Notes

1. All direct quotations in this section are taken from published accounts in *The New York Times*, May 9–11, 2006.
2. Regional media outlets quickly picked up the story of President Ahmadinejad's letter to President Bush. A narrow search on *Lexis-Nexis* revealed 230+ stories from May 9–May 14 from "Middle East / African Sources" in *Universe's* "World News" library. Searching the "European News Sources" showed 120+ hits, 79 in the "North / South American News Sources" library, and 120 hits in the "Asian Pacific News Sources" library. Though not identifying duplicate records, the narrow search ("Bush" w/10 "Iran" and "letter" from "05/09/2006" to "05/14/2006") revealed sources dealing primarily with reaction to the letter, rather than more general stories about U.S.–Iranian relations. Collectively, those four libraries represent over 1,000 regional newspapers, journals, and wire services.

3. Or, in fairness, it could be a political feign, similar to the bogus invitation to dialogue supposedly aimed at achieving peace offered by the North Vietnamese to President Nixon during the Vietnam War. The result was a prolonged and frustrating exchange of charges and counter-charges organized around the idea of "Paris Peace Talks."

References

AFX International Focus. (2006, May 10). Letter from Iran does not address nuclear worries: Bush. Retrieved May 14, 2006, from LexisNexis Academic Database.

Agence France Presse. (2006, May 10). Iran's letter 'not a serious diplomatic overture': Rice. Retrieved May 14, 2006, from LexisNexis Academic Database.

Al-Issawi, Tarek. (2006, May 11). High-ranking Iranian representative offers new suggestions to resolve nuclear crisis. Associated Press Worldstream, Retrieved May 14, 2006, from LexisNexis Academic Database.

BBC Monitoring Middle East. (2006a, May 14). U.S. silence towards Iran's letter illogical. Retrieved May 14, 2006, from LexisNexis Academic Database.

BBC Monitoring Middle East. . (2006b, May 12). Top cleric in Iran's Khuzestan comments on president's letter to George Bush. Retrieved May 14, 2006, from LexisNexis Academic Database.

Corman, S. R., Kuhn, T., McPhee, R., and K. Dooley (2002). Studying complex discursive systems: Centering resonance analysis of communication. *Human Communication Research*, 28 (2), 157–206.

Cornwell, Rupert. (2006, May 10). Iran's letter to America cuts no ice in nuclear crisis. *The Independent* (London), p. 34, Retrieved May 14, 2006, from LexisNexis Academic Database.

Deutsche Presse-Agentur. (2006, May 11). Ahmadinejad: I called on Bush to return to spirituality. Retrieved May 14, 2006, from LexisNexis Academic Database.

Eisenberg, E., Goodall, H. L., & Trethewey, A. (2006). *Organizational Communication: Balancing Creativity and Change,* 5th ed. New York: Bedford/St. Martin's.

Emirates News Agency. (2006, May 10). Iran, U.S. starting from scratch. Retrieved May 14, 2006, from LexisNexis Academic Database.

Freeman, R. Edward. (1984). *Strategic management: A stakeholder approach*. Boston: Pitman.

Kabalan, A. M. U. S. will not allow a nuclear Iran. *Gulfnews.com.* Retrieved May, 14, 2006 from http://archive.gulfnews.com/articles/06/05/12/10039277.htmlgng.

Kole, William. (2006, May 11). World wonders: Why won't U.S., Iran hash out differences face to face? *Associated Press Worldstream,* Retrieved May 14, 2006, from LexisNexis Academic Database.

La Guardia, Anton. (2006, May 9). Iran's letter 'a ploy to avert pressure.' *The Daily*

Telegraph (London), p. 14, Retrieved May 14, 2006, from LexisNexis Academic Database.

Lewis, Bernard. (1988). *The political language of Islam*. Chicago : University of Chicago Press.

Mideast Mirror. (2006, May 10). Reading between the lines. Retrieved May 14, 2006, from LexisNexis Academic Database.

Mitchell, Ronald K., Agle, Bradley R., & Wood, Donna J. (1997). Toward a theory of stakeholder identification and salience: Defining the principle of who and what really counts. *Academy of Management Review, 22*, 853–886.

The White House (2006, May 7). Interview of the President by Kai Diekmann of BILD. Retrieved May 14, 2006 from *http://www.whitehouse.gov/news/releases/2006/05/20060507-2.html*

Taheri, Amir. (2006, May 14). Reading between the lines. *The Jerusalem Post*, p. 13. Retrieved May 14, 2006, from LexisNexis Academic Database.

Wright, Robert (2006, May 14). They hate us, they really hate us. *The New York Times* online.

Xinhua General News Service. (2006, May 11). U.S. rebuffs direct talks with Iran. Retrieved May 14, 2006, from LexisNexis Academic Database.

CHAPTER SIX

One Message for Many Audiences

Framing the Death of Abu Musab al-Zarqawi

Z. S. JUSTUS AND AARON HESS

Introduction

Globalization and telecommunications technology have made every message global. The consequence of this phenomenon is that when the United States makes announcements concerning the Global War on Terrorism a global, rather than a local, audience receives the message. While similar messages may circulate in different areas throughout the globe, the messages interact with national and/or cultural traditions that result in different types of message interpretation.

Using Entman's (2003a; 2003b) Cascading Network Activation model we charted the news of Zarqawi's death, specifically the photographs of his death, as those messages reached different audiences. The news was interpreted in a wide variety of ways by a great number of actors. For example, Jihadi leadership and media moved quickly to proclaim him as a martyr through the reframing of the photographs of his body in an attempt to continue their campaign of terror. In this analysis, we trace the story of Zarqawi's death through three media outlets: the mainstream United States press, Al-Jazeera, and Jihadi media.

We offer a modification of the Entman model and three policy recommendations to assist in adapting messages for a global audience.

First, we urge policy makers to accept that messages are global. Policy makers must consider if/how messages and images could be re-appropriated by individuals who hold an antagonistic stance toward the United States. Our second major policy recommendation reflects the harsh reality of a global media environment. Policy makers should immediately adopt a formal pre-release analysis and decision-making process which takes into account the likely effect of those messages on strategic cultures such as those in the Middle East, in addition to the effect of those messages on domestic audiences. Finally, we offer a suggestion based on the specific issue of publicizing death photos of Jihad leaders. Decision makers should seriously consider (1) preventing images of the dead from mass circulation, (2) releasing information about the dead from non-US sources, and (3) avoiding messages that portray the United States as voyeuristic and/or barbaric.

The adoption of these three message strategies provides a framework that accurately reflects the reality of global communications. In addition, the third recommendation will have a positive impact in the specific scenario of releasing information concerning dead Jihadi leaders. Together, these policy recommendations represent general principles and a specific application of those principles that will help the United States win the Global War on Terrorism.

Background

The United States does not normally publish photographs of citizens or soliders who die in war. There have been exceptions. For example, in response to questions concerning the publication of the death photos of Uday and Qusay Hussein, Secretary Rumsfeld commented "It is not a practice the United States engages in on a normal basis" (Hedges, 2003, p. A23). Publication and mass circulation of death images are a complicated subject that some critics believe is covered under the Geneva Convention code prohibiting the publication of photographs of prisoners of war. Others advocate the practice when it demonstrates the loss of a major political figure or terrorist leader.

Despite the general prohibition on publication of photos of war casualties the United States elected to make public certain photos in the Global War on Terrorism. The first notable exception occurred when the Department of Defense allowed Uday and Qusay Hussein to be photographed in July 2003.

According to government officials the publication of the photographs was justified because "there was no other, less graphic, way to prove to people that the potential heirs of Saddam's Baathist regime were gone" (Hedges, 2003, p. A23). In addition, the photos were said to provide "higher troop morale, more intelligence from Iraqi people and irrefutable evidence that Saddam's tyranny is over" (Manly, 2003, p. 4).

This highly controversial media event served as a preview of the more recent publication of photos of Abu Musab al-Zarqawi. On June 8th, 2006, the US military killed al-Zarqawi in a coordinated coalition effort. Locating Zarqawi in a safe house near the town of Baqubah, the U. S. Air Force dropped two 500-lb. bombs on the house. Following the strike, Iraqi Police discovered Zarqawi, who survived for an additional 52 minutes. His death was seen as a success for the Bush Administration and for the war effort in Iraq because one of the terrorist masterminds and leaders had been eliminated and additional leads on al-Qaeda in Iraq were found in a "treasure trove" of information amongst the rubble. In announcing the death of Zarqawi, Army Major General William B. Caldwell led a press conference detailing the strike and subsequent identification of the body. In doing so, Caldwell displayed maps of the area where the safe house was located, video of the air strike, and images of the deceased Zarqawi.

The Media Response

In reaction to the display of Zarqawi's body, various news outlets noted that images of his death were "gruesome" (Adams, 2006, p. 12), a "trophy" (Kennicott, 2006, p. C01), and even "sanitised" (Nason, 2006, p. 10). The mixed responses highlight the complexity of the global media stage, where information and images travel faster than ever. Editors of newspapers received complaints of the graphic nature of the images appearing on front pages, arguing that the display was expected from the terrorist enemy, not from the United States (Diadiun, 2006). In descriptions of the trophy-like status of the document, writers note that its presentation in a professional mat and frame the photograph looked as "something one might preserve and hang on the wall next to other family portraits" (Kennicott, 2006, p. C01). Finally, the sanitization of the photograph was noted through Caldwell's statements that the body was altered before the image was obtained. Debris and blood were removed to make the body more presentable and in hopes of not inflaming the Islamic World (Nason, 2006). Given the mixed reviews of the photographic evidence of Zarqawi's death, questions should be raised regarding the efficacy of such a display, especially in the environment of global media.

In examining the prominent news outlets, the story was disseminated across the globe in a matter of minutes after the initial press conference. Al-Jazeera listed Zarqawi's obituary on the same day, remarking about the exaggeration of his prominence ("Obituary," 2006). On June 12th, four days after the strike, Al-Jazeera featured a photograph of the news conference including the image of Zarqawi's body ("Doubts shroud," 2006). Bloggers also tracked the story as it progressed. From both sides of the political spectrum ("georgia10" at *Daily Kos* and Michelle Malkin's Blog), the story broke the same day with fellow bloggers adding to the general commentary. With a global network of information flow, the dissemination of details regarding the event moved at a lightning-fast pace, which means that words and images must be carefully crafted to avoid cultural insensitivity and misinformation.

In examining the prominent news outlets, the story was disseminated across the globe in a matter of minutes after the initial press conference. Al-Jazeera listed Zarqawi's obituary on the same day.

Given the policy of photographic censorship and its rare exceptions, the notable event of the death of Zarqawi and the subsequent mass circulation of images of his body were received with mixed results. Importantly, letters to the editor and other news outlets questioned the use of the image due to its graphic and trophy-like status. While the U. S. military took cautionary steps in the production of the image through cleaning the body, the image and the means of execution still struck many as barbaric ("Talk back to the media," 2006; "Allies wrong to display," 2006; Diadiun, 2006). Increased global circulation of images gave them political weight and communicative importance, especially in a global news cycle. In short, this death image was on display, with commentary, everywhere.

We offer the following analysis. First, we discuss the nature of the global audience vs. local audience. With the use of mass media outlets, the nature of audiences and their reception to messages have changed over the past 20 years. Second, we examine scholarship which has approached the problem of how messages are interpreted by global audiences in various sectors of society, including business, politics, and culture. Finally, we review the work of visual communication and rhetoric scholars who have evaluated the impact of images on diverse audiences. Our aim in this chapter is to illuminate the possible effects of image distribution within global networks.

The Power of Globalization

Globalization has profound implications for the way information is distributed. Smith (1999) noted that control over information is "passing into the control of managements (of that information) whose outlook is exclusively global" (p. 355). This change in focus is a reflection of how information circulates in a media-saturated global news environment. Put another way, there is no such thing as a local message. Ward (2005) writes:

> News reports, via satellite or the Internet, reach people around the world and influence the actions of governments, militaries, humanitarian agencies, and warring ethnic groups. The reach of the Al-Jazeera and CNN networks, for example, extends beyond the Arab world or the American public. (p. 4)

Not only are media outlets global in scope, but they are also interconnected and overlapping. This produces a remarkably fluid and rapid dissemination of information that crosses national and cultural boundaries with little impediment. In addition, "national governments cannot easily enforce even the modest rules some have adopted to regulate or impede these processes" (Smith, 1999, p. 356).

"News reports, via satellite or the Internet, reach people around the world and influence the actions of governments, militaries, humanitarian agencies, and warring ethnic groups" (Ward, 2005, p. 4).

The issue that both Ward and Smith hint at is that while messages are primarily universal, audiences are not. Ward's (2005) commentary on Al-Jazeera's reach beyond the Arab world is especially interesting. Al-Jazeera is accessible on the Internet, and, more importantly it is accessible in English. The problem we are left to grapple with is one of a single message reaching multiple national and cultural audiences.

A Framework for Message Dissemination

Robert Entman (2003a, 2003b) proposes a model for understanding how information is disseminated from government sources to the public. Entman's work is exceptional for two reasons. First, his model has been adopted and/or cited my numerous researchers as an accurate framework for evaluating message framing and dissemination (Byerly, 2005; Cherribi, 2006; Dahinden, 2005; Jerit, 2005; Ross & Bantimaroudis, 2006; Tilley & Cokley, 2005; Wang,

2006; Wicks, 2005). Second, the model itself is exceptional in that it not only isolates key stages of message development but also provides insight into who has influence over the message as it changes and moves.

Figure 1. Cascading Network Activation Model (Entman, 2003a, p. 419

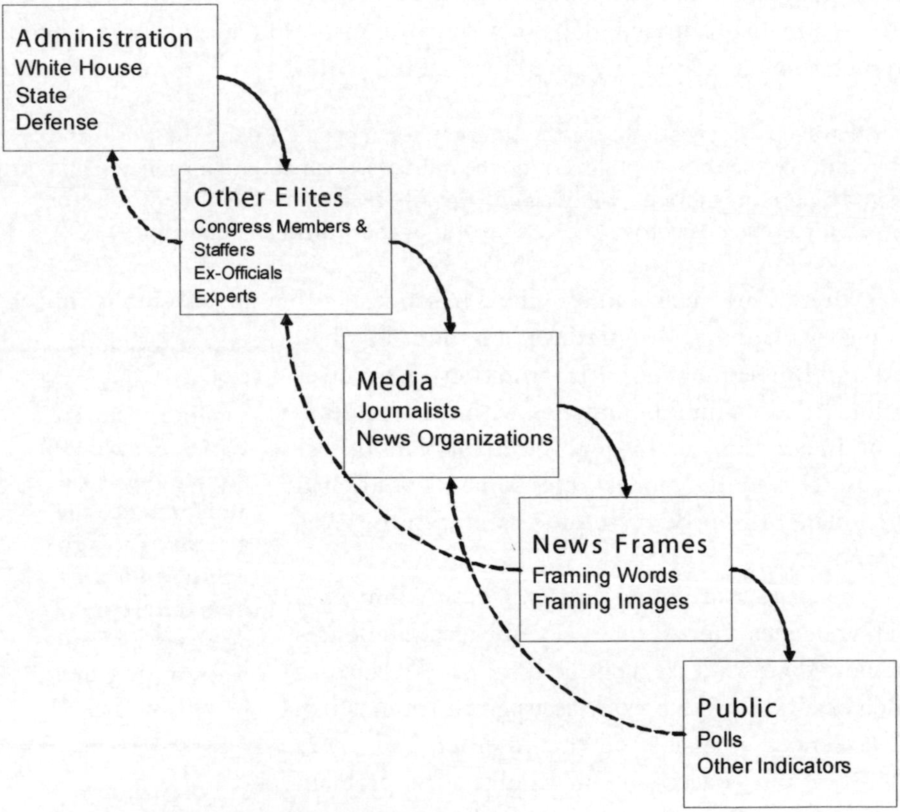

Administration
White House
State
Defense

Other Elites
Congress Members &
Staffers
Ex-Officials
Experts

Media
Journalists
News Organizations

News Frames
Framing Words
Framing Images

Public
Polls
Other Indicators

In this model:

- Administration/Military: At this level command decisions are made about how to deal with events.
    - President Bush and Secretary Rumsfeld take advice from staffers and make the final call about how to make information public.
- Other Elites: Staffers, such as press secretaries, convey the information

and message agenda from the Administration and communicate with the press.

- These elite information gatekeepers maintain a close eye on the press and can adapt, or spin, their message accordingly.

- Media: Members of the media receive information from staffers and other information elites. The media keep tabs on polling data from the public as it attempts to craft a more marketable message.
- News Frames: This area represents the media's crafting of a specific message.
    - News frames emphasize certain details and omit others in an effort to mold information into a more acceptable (to the elite stakeholders') story.
- Public: The public receives information from press sources and forms opinions based on it and in response to it.

In short, this model provides a way to understand the stages a message goes through (i.e., "cascades") as it makes its way from decision makers to the public. An important aspect of this cascading model is its *portability*. Cherribi (2006) notes:

> The metaphorical model of the cascade created by Robert M. Entman in his book *Projections of Power* (2003b) to explain the process of influence over the frames projected into the news about foreign crises, public opinion, and elite thinking offers a dynamic approach to analyze a global media outlet like Al-Jazeera. (p. 134)

Cherribi (2006) focuses his discussion on the political controversy over the use of veils among Muslim women in France. He uses Entman's model and concludes "Al-Jazeera may, on the surface, look as if it offers pluralism with its variety of programs and opinions. In the case of the veil, however, there is only one perspective, an Islamic perspective that is to encourage women to wear the veil" (p. 134). The important aspect of this study is that it displays the utility of Entman's model (2003a; 2003b) in the context of the mass circulation of political messages throughout the Middle East.

The diagram used above may be modified and expanded to enrich our understanding of how single messages, such as the death of Zarqawi, can be locally appropriated and re-interpreted by multiple audiences. Returning to the theme from earlier—*there are no local messages*. We adopt a modified version of Cherribi's (2006) Al-Jazeera diagram to reflect the release of information about Zarqawi's death and provide a model of how information flows within each media network. Compared to the traditional cascading model proposed by Entman (2003a; 2003b) there are many opportunities for constructing more appropriate messages for international audiences.

Figure 2. Global Cascading Activation Networks Model

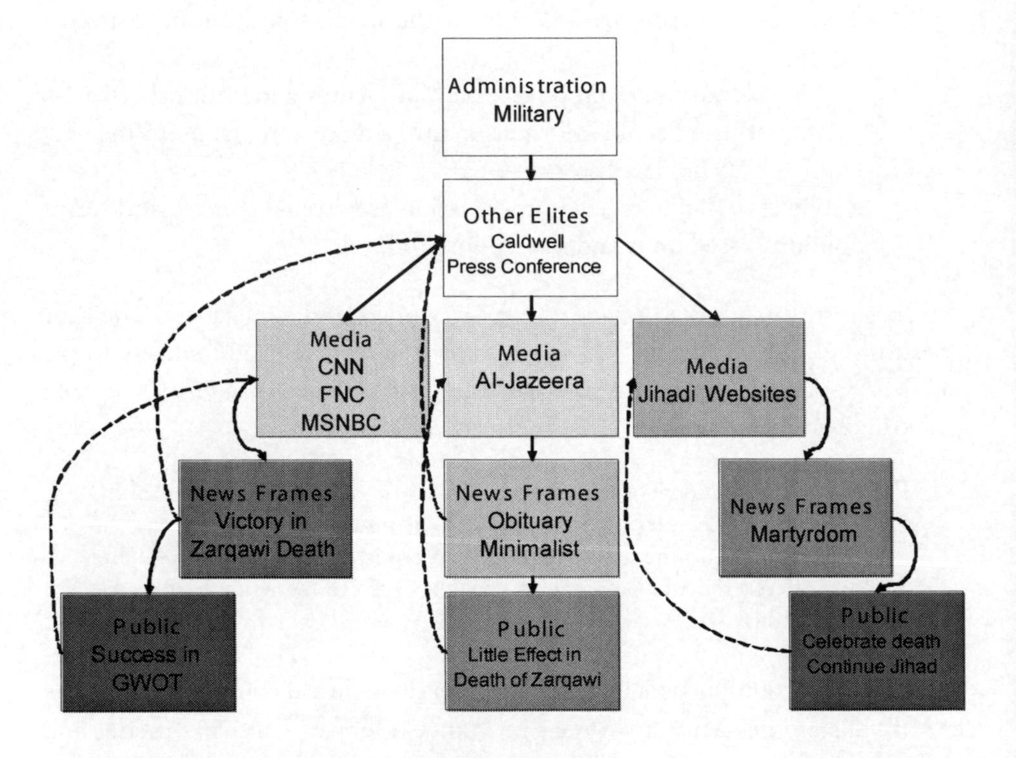

The right side of the diagram that outlines the distribution of information along Jihadi channels deserves further explanation:

- The first and second levels of the diagram remain the same. Administration officials and staffers still have a degree of control over the information that becomes public.
- At the third level, Jihadi websites are a primary mechanism of information distribution (Corman & Schiefelbein, 2006). It should be noted that websites/bloggers also obtain information from Al-Jazeera and from Western news sources.
    - These websites also report on and react to the actions and be-

liefs of a public that is sympathetic to Jihad.

- The websites/bloggers report the news in a differing ways; but more importantly, how they see and interpret the information is beyond the control of Western media elites and their spokespersons. In the case of the circulation of the death images of Zarqawi, *it was the release and circulation of those images that led to his death being framed as worthy of martyrdom.*
- Diverse publics receive and likely adopt the interpretation of the more culturally and politically aligned websites. Zarqawi's death is both mourned and celebrated as an act of martyrdom.

Asia Times reporter Michael Scheuer (2006) writes:

> On bin Laden's side, al-Qaeda publicly will mourn Zarqawi's death, recall him as a noble and selfless mujahid, and cite him as a brave comrade-in-arms killed by the crusaders' high-tech aircraft while he was armed only with faith and an AK-47. This is likely the way many Muslims outside Iraq recall him, *thanks in large measure to the post-attack photograph US public relations officers distributed of Zarqawi's face*. (n. p.; emphasis ours)

Visuals, especially photographs, are very powerful symbolic forms of communication and influence (Finnegan 2006; Lucaities and Hariman, 2001; Scott, 2004). The photographic information presented in a military press conference is the exact same photographic information that appears on a Jihadi website, but the interpretation of its meaning is entirely dependent upon the local framing of the image within a specific cultural and political environment. Our recommendations for navigating this situation are the subject of the following section.

Three Guidelines for Message Creation in a Global Context

1. **Recognize that messages are global**: Information distributed at a military press conference is available to domestic media outlets, Al-Jazeera, conservative and liberal blogs, and jihadi websites, all in a matter of hours.
    a. When deploying messages concerning new events in the

Global War on Terror, consider the impact of various media organizations and their ability to reframe messages in reference to their cultural politics. Use the above revised Entman model to predict responses to global messages.

b. Absolute control over the message is impossible. Craft messages that result in meaningful exchange. Constructing ambiguous messages can convey appropriate information as well as generate dialogue (Goodall, Trethewey & McDonald 2006).

Guidelines:
1. Recognize that messages are global.
2. Adopt an independent analysis and message coordination process.
3. Reconsider the release of images of dead leaders.

2. **Adopt an independent analysis and message coordination process**: The unfortunate reality of global media is that it is impossible to please everyone with one message due to different religious/cultural/national circumstances. Situations will present themselves when the release of information will please one group but cause others to grow uneasy. We suggest that in these unfortunate circumstances the government should adopt a consistent media strategy that focuses on the needs of strategic cultures (such as the Middle East) rather than other options.
3. **Reconsider the release of images of dead leaders:** Photos of the death of Zarqawi are seen as signs of progress in the GWOT for domestic audiences, but the larger global audience did not interpret them in the same way. Part of recognizing the reality of global media is accepting that messages must be crafted for a global audience, as well as a domestic one, and that the messages, media strategies, and reinforcement mechanisms are likely to require far more complexity and subtlety than is currently in place.
   a. Due to the current low credibility status of the United States government (Corman, Hess, and Justus, 2006) *utilize third-party credible sources* to release controversial information and images and to confirm the validity of the diplomatic and military acts used to obtain them. In the Zarqawi example, it may have been more politically useful for the newly elected Iraqi government officials to release the story of his death.
   b. Avoid the perception that the United States engages in barbaric

practices. The display of Zarqawi's body has been compared to the beheading videos which he orchestrated and which make the U.S. military actions symbolically equivalent to those barbaric acts. *In all things symbolic, avoid media tactics that allow the comparison between Jihadi organizations and the US government.*

c. If and when Osama bin Laden is located, if he cannot be captured alive, avoid symbolic displays of his death which support his ascension to martyrdom. If he is found alive, use a more culturally sensitive approach to releasing the story and any accompanying photos to demystify his status as the leader of al-Qaeda by publicizing his accountability to the international justice system.

Conclusion

The recent controversies over the display and mass circulation of the death images of al-Zarqawi highlight the need for a more culturally sensitive and symbolically nuanced approach to news releases. Jihadi leaders and media quickly moved to proclaim his martyrdom and, in turn, mitigated the death of Zarqawi as a victory in the war on terror. In the event of another leader's capture, following the above guidelines will prevent such a hasty reappropriation of our success.

Our adaptation of Entman's model of cascading influence provides a way of thinking about the diverse ways in which messages are reframed by media outlets and audiences. This model also strongly suggests the necessity of creating an independent mechanism for the pre-release cultural and political analysis of the likely effects of messages on diverse media markets and audiences.

References

Adams, P. (2006, June 13). Zarqawi killing not the answer. *The Australian,* p. 12.

"Allies wrong to display trophy photograph of al-Zarqawi" (2006, June 12). *The Daily Telegraph (London),* p. 21.

Byerly, C. (2005). The formation of an oppositional discourse. *Feminist Media Studies, 5,* 281–296.

Cherribi, S. (2006). From Baghdad to Paris: Al-Jazeera and the veil. *Harvard International Journal of Press/Politics, 11,* 121–138.

Corman, S., Hess, A. & Justus, Z. S. (2006). *Credibility in the Global War on Terrorism: Strategic Principles and Research Agenda.* White Paper #0603, Consortium for Strategic Communication. Available online: *http://www.asu.edu/clas/communication/about/csc/documents/csc_credibility_gwot.pdf*

Corman, S. & Schiefelbein, J. (2006). *Communication and media strategy in the Jihadi war of ideas*. White Paper #0601, Consortium for Strategic Communication. Available online: *http://www.asu.edu/clas/communication/about/csc/publications/jihad_comm_media.pdf*

Dahinden, U. (2005). Framing: A decade of research experience. *Conference Papers—International Communication Association*.

Diadiun, T. (2006, June 11). Readers could have done without photo of dead terrorist. *Plain Dealer,* p. A2.

"Doubts shroud Zarqawi's last minutes" (2006, June 12). *Aljazeera.net.* Retrieved June 22, 2006 from *http://english.aljazeera.net/NR/exeres/C76601D9-F56C-4732-8A86-FDE8D867C7CD.htm*.

Entman, R. M. (2003a). Cascading activation: Contesting the White House's frame after 9/11. *Political Communication*, *20*, 415–433.

Entman, R. M. (2003b). *Projections of power: Framing news, public opinion, and U.S. foreign policy*. Chicago: University of Chicago Press.

Finnegan, C. (2006). What is this a picture of?: Some thoughts on images and archives. *Rhetoric & Public Affairs*, *9*, 116–123.

georgia10 (2006, June 8). Abu Musab al-Zarqawi is dead. *Daily Kos.* Retrieved June 22, 2006 from *http://www.dailykos.com/storyonly/2006/6/8/75854/67368*.

Goodall, B., Trethewey, A., & McDonald, K. (2006). *Strategic ambiguity, communication, and public diplomacy in an uncertain world: Principles and practices.* White Paper #0604, Consortium for Strategic Communication. Available online: *http://www.asu.edu/clas/communication/about/csc/documents/StrategicAmbiguity.pdf*

Hedges, M. (2003). Photos of brothers' bodies released; U.S. shows images of Uday, Qusay, Hussein as proof, but some fear backlash in Iraq. *The Houston Chronicle,* p. A23.

Jerit, J. (2006). Reform, rescue, or run out of money? Problem definition in the social security reform debate. *Harvard International Journal of Press/Politics*, *11*(1), 9–28.

Kennicott, P. (2006, June 9). A chilling portrait, unsuitably framed. *The Washington Post,* p. C01.

Lucaites, J., & Hariman, R. (2001). Visual rhetoric, photojournalism, and democratic public culture. *Rhetoric Review*, *20*, 37–42.

Malkin, M. (2006, June 8). Zarqawi has been killed. *Michelle Malkin*. Retrieved June 22, 2006 from *http://michellemalkin.com/archives/005351.htm*.

Manly, H. (2003, July 27). Trusting Bush a causality of war. *Boston Herald,* p. 4.

Nason, D. (2006, June 10). Clean hit on a dirty fighter. *Weekend Australian,* p. 10.

"Obituary: Abu Musab al-Zarqawi" (2006, June 8). *Aljazeera.net.* Retrieved June 22, 2006 from *http://english.aljazeera.net/NR/exeres/128D6EEE-8BCA-46CB-AF05–417E14D40AEA.htm.*

Ross, S., & Bantimaroudis, P. (2006). Frame shifts and catastrophic events: The attacks of September 11, 2001, and *New York Times*'s portrayals of Arafat and Sharon. *Mass Communication & Society, 9*, 85–101.

Scheuer, M. (2006, Jun 22). Zarqawi's death an opportunity for al-Qaeda. *Asia Times.* Retrieved June 22, 2006 from *http://atimes.com/atimes/Middle_East/HF22Ak01.html.*

Scott, B. (2004). Picturing irony: the subversive power of photography. *Visual Communication, 3*, 31–59.

Smith, A. (1999). Media globalism in the age of consumer society. In J. Hanson & Maxcy, D. J. (Eds). *Sources: Notable selections in mass media* (354–363). Guilford, CT: Dushkin/McGraw-Hill.

"Talk back to the media" (2006, June 17). *Rocky Mountain News,* p. 12C.

Tilley, E., & Cokley, J. (2005). Unfreezing the frame: Functional frame analysis for journalists and journalism students. *Australian Journal of Communication, 32*, 71–88.

Wang, J. (2006). The politics of goods: A case study of consumer nationalism and media discourse in contemporary China. *Asian Journal of Communication, 16*, 187–206.

Ward, S. J. A. (2005). Philosophical foundations for global journalism ethics. *Journal of Mass Media Ethics,* 20, 3–21.

Wicks, R. (2005). Message framing and constructing meaning: An emerging paradigm in mass communication research. *Communication Yearbook, 29*, 333–361.

CHAPTER SEVEN

Re-Defining the Long War

Toward a New Vocabulary of International Terrorism

AARON HESS AND Z. S. JUSTUS
WITH CONTRIBUTIONS FROM: KRIS ACHESON AND STEVEN R. CORMAN

Introduction

As the fight against terrorism continues, language plays a pivotal role. In current policies, the language of war continues to dominate. Based on an analysis of President Bush's September 11th anniversary campaign speeches, we propose that war metaphors and language, such as victory, enemies, and allies, occlude the reality of counterterrorism efforts. It is difficult to pinpoint victory in this conflict, a requisite of the vocabulary of war familiar to lay audiences.

We call for a new language to illustrate the nature of our present conflict, a new vocabulary of international crime as an effective replacement for discussions of counterterrorism. There are four benefits of this new language. First, domestic audiences are accustomed to the persistence of crime; it is a manageable social ill. Second, the labeling of terrorist organizations as "criminal" decreases the perceived legitimacy of their acts by potential recruits. Third, international crime is a global problem, not a war perpetrated by the United States. Global problems require global solutions, and such a language will help garner support from the global community. Finally, crime language separates the religious connotation associated with labels of terrorism or "jihadism." This allows moderate Muslims to reframe their faith away from extremist and violent acts.

To illustrate the new language, we have modified one of President Bush's speeches to remove uses of war language and replace it with a new language of international crime. The speech reads just as "tough" on terrorism while avoiding the disadvantages of war framing. Future domestic and international persuasive efforts to win support should take into account this new language of international crime.

Background

The level of popular domestic support for the Global War on Terrorism (GWOT) is in decline. According to a recent poll by Rasmussen less than half of the American population believes we are winning the Global War on Terrorism.[1] While there are a number of possible explanations for flagging support of the GWOT, it is worth examining the relationship between the perception of victory and support. In the current language of the GWOT, Americans hear of a war being waged upon an enemy. However, communicating progress of such a war is difficult due to the non-traditional nature of the GWOT. Simply, the rhetorical framework of the GWOT does not fit the reality of the conflict.

The Global War on Terrorism is different than any war that has ever been fought. We are constantly reminded of this fact from a number of different sources. The 2006 National Security Strategy noted, "Our strategy also recognizes that the War on Terror is a different kind of war."[2] The Permanent Select Committee on Intelligence more specifically commented that "To win the war on terrorism, the United States and our allies will have to not just kill and capture key terrorist operatives."[3] The 2002 National Security Strategy as well as President Bush and members of his cabinet all agree that this is a war unlike any war we have ever fought.

"European leaders believe that the language of war inflates those who are in fact terrible criminals."

Mary Robinson -American Bar Association

We agree that the war on terrorism is different—so different in fact that it is not a "war" at all. This conflict features non-state actors who use methods that violate international standards of war. They have no standing army and it is unlikely that the conflict will end in a peace accord. Additionally, in recent months, criticism of the phrase "war on terror" has grown. The British government has elected to not use the word in future operations[4] and has criticized the United States' preference for a war vocabulary.[5]

Several commentators have openly criticized the language choices made by the Bush Administration in describing the GWOT. Senator Russ Feingold questions the accuracy and strategic value of the term "Islamic fascist," a rhetorical effort to recall World War II enemies. "Call them whatever you want—monsters, butchers—but the use of the term 'Islamic fascist' puts the name of Islam . . . in an exceptionally negative light."[6] Mary Robinson of the American Bar Association recognizes the international nature of the debate.

European leaders believe that the language of war inflates those who are in fact terrible criminals, through committing acts of terrorism that deliberately kill civilians, and that the context of being "at war" makes it more acceptable to erode standards of civil liberties and human rights.[7]

Without the proper vocabulary, it is impossible to accurately describe the conflict. We believe that communication scholarship can add a level of depth and sophistication to these arguments to fully explain the problems of the current vocabulary of war. In the following section we will explain the importance of the label "war" within the Global War on Terror by reviewing communication scholarship concerning the rhetoric of war. We also evaluate how the vocabulary used to describe armed conflict has developed and been used in the last sixty years and its shortcomings in describing the current conflict.

The Recent History of Naming War

As the world emerged from the horror of World War II and entered the nuclear age, one thing did not change—the language of war. In point of fact, the language of war has changed very little for several hundred years. We can easily use *victory* to describe military engagements as diverse as the English defeat of the Spanish armada and Allied victory at Iwo Jima. Similarly concepts such as *war*, *defeat*, *offensive*, *attack*, and *surrender* have changed very little over time. Even as war turned metaphorical during the Cold War the vocabulary stayed the same. Medhurst's observation helps to explain this stasis. He noted that the "Cold War, like its 'hot' counterpart, is a contest." This striking similarity is more than

American presidents regularly invoke the vocabulary of war to communicate important messages to their constituents.

a casual observation. Even though the Cold War was fought with "markets, spheres of influence, and military alliances, as well as such intangibles as public opinion, attitudes, images, expectations, and beliefs"[8] instead of guns and rockets, the language used to describe the contest remained the same.

American presidents regularly invoke the vocabulary of war to communicate important messages to their constituents. Robert Ivie undertook an analysis of 150 years of war rhetoric in American presidencies, that "these several war messages establish a pattern" leading to a recognizable linguistic framework "through which American Presidents assess international relations." Ivie continues, arguing that the rigid vocabulary of war has held steadfast, "despite major variations in the nature of the enemy, the intensity of the threat, the personality of the presidents, (and) the general historical milieu." Following this vocabulary "produces (a) set of ideal purposes and agencies that, presumably, should be adhered to by all men in all places."[9] The language of war is relatively stable; it communicates a set of conditions, including moral justifications, conditions for victory, and the nature of the enemy. By using scholarship about the historical vocabulary of war, our analysis tracks President Bush's adoption of the war framework to communicate the GWOT, despite the overwhelming dissimilarity between the current struggle and traditional concepts of war.

Analysis

In an effort to understand how the language of war pervades rhetoric about the GWOT we have undertaken an extensive analysis of a series of five speeches that President Bush gave between August 31, 2006 and September 29, 2006.[10] These speeches were part of a series intended to build support for the War on Terror around the five-year anniversary of 9/11. These speeches represent the administration's best efforts to reframe the war in a more positive light despite recent setbacks. A number of commentators have characterized these speeches as "blunt," indicating that not only are these speeches a best effort, they also leave very little to the imagination.[11] We analyzed the speeches by parsing words and phrases that describe the enemies, allies, and outcomes/events in the GWOT. Table 1 on page 134 shows the short words and phrases extracted from the speeches about the war. It is organized by the following narrative elements: people, actions, means, goals, and environment.

Our analysis reveals that the speeches rely on very traditional language and description in their characterizations of the GWOT. Obviously, the term "war" is the central element in the acronym GWOT. But the presence of other language also points toward a very traditional conceptualization of the current fight against terrorism. In the narrative contained within the speeches: "allies" form a "coalition" that uses a "clear plan" to stay on the "offensive" in order to "defend civilization" and "win the war" all aimed at the end goal of "complete victory" within a "global campaign." This narrative is juxtaposed with that of the terrorists, which characterizes them as "enemies" and their "terrorist allies" who utilize "propaganda," "strategy," and pursue "weapons of mass destruction" as part of a larger effort to "defeat" our troops through "multiple attacks" with the final goal of creating a "violent political utopia" and "forcing America to retreat."

The above language cannot be described as inaccurate. Many parts of the descriptions ring true and seem to follow what we and other Americans see in the news and believe is happening. However, war language is a choice from among other possibilities, one that has rhetorical consequences that can diminish long-term chances of success, and the war framing has fundamental problems.

For example, using traditional language to describe the war on terrorism leads people to believe that this war is similar to other wars. This leads the public to expect victory, something that can be "complete" and result in the placement of an American flag on a hill. Indeed, Bush frequently references other victories in American history, including World War II. While previous conflicts have had "V-days" that symbolized the finality of a conflict, the GWOT may never have a day where conflict fully subsides. Even when the coalition performs a successful operation, because of the reliance on traditional war language, the president is unable to communicate the magnitude of the success accurately to the American public, yet such progress reports are expected under a traditional war framing.

A striking example of this problem is found in the recent death of a terrorist leader. The death of Abu Musab al-Zarqawi was a major success for the coalition in Iraq. However, the display of his corpse paved his path to martyrdom among sympathizers and met a lukewarm reception in the United States.[12] Commanders knew that his death was important, but because the reality of this conflict is different, the event could not be accurately translated into

While previous conflicts have had "V-days" that symbolized the finality of a conflict, the GWOT may never have a day where conflict fully subsides.

Table 1. Existing Characterizations of the Global War on Terrorism

	US	THEM
Action	Defending civilization, Killed hundreds of Taliban fighters, Fight the enemy, Foolish to negotiate, Memorial(ize), (We will not) Retreat, Not give the enemy victory, Winning this war, Stay on the offense, Taking the battle (to them), Given their lives, On the offense	Attacked, Murdered, Massed fighters, Fight our forces, Massacre, Defeat and disgrace, Defeat, Localized strike, Multiple attacks
Goals	Victory, Liberated, Bring you to justice, Triumph, Complete victory, Win, Eliminate extremism, Defeat al Qaeda, Completing the mission, Win this war on terror, Emerge victorious, Liberty triumph, Win the ideological struggle, Prevail	Violent political utopia, A totalitarian nightmare, Radical Islamic empire, Radical dictators, No compromise or dialogue, Forcing America to retreat, Regime of tyranny, Open field, To control governments, Terrorist states, Armed with nuclear weapons, Decentralized operating bases, Totalitarian empire
People	Friends and allies, Strong ally, Coalition, Forces of freedom, Allies, Allies in the war on terror, Nation at war	Enemies, Tyrants, Radical, Dictators, Enemy we face, Mind of our enemies, Terrorist allies, Enemies, Enemies of liberty
Means	Long war, Clear plan, Fierce fighting, Arsenal, Offensive, Military might	Resources, Nuclear, Weapons, Propaganda, Strategy, Strategy to defeat us
Environment	Global network, Empire, Conflict, War, Struggle, Across two oceans, Military conflict, Central front, Third World War, Greatest battle, Pivotal moment, Despotism, Battlefront, Wars of the 20th century, Path to victory, War of destiny, Central battlefield, Global campaign, Campaign against terror	

the language of traditional warfare. The death of Zarqawi was not "complete victory," but it was important. Yet, it was impossible to convey its significance within the traditional framework.

The inability to communicate victory using the traditional language of war is directly related to flagging support for the war on terrorism. As the aforementioned scholars argue, perceptions of victory are closely tied with support. In short, the inadequacy of traditional war framing is, at least partially, to blame for a lack of domestic enthusiasm regarding the GWOT. The solution to this problem is more complicated than renaming the GWOT as the "global struggle against extremism"—we need a new, more accurate vocabulary for the entire conflict. It must communicate the conflict accurately, contain an appropriate tone, and frame the conflict in terms compatible with its likely path of development. In crafting the vocabulary we must be aware that while the problem we have identified is domestic, everyone will have access to the words we choose. While it is impossible to craft the perfect message/vocabulary for every audience,[13] we believe that we can make substantial improvements on the current language choices we have made.

Policy Recommendations

We propose that the United States Federal Government should adopt a new vocabulary in reference to what is now called the Global War on Terrorism. In discussing the debate over the language of war, Walid Phares[14] calls for a clear alternative to present language, an alternative that presents the conflict as both ideological and global. We answer that call by providing a framework that confers international legitimacy to counterterrorism efforts as well as belittling the motivations for violent acts by groups such as al-Qaeda.

Phares warns that to re-label the war would to embolden terrorist groups by not recognizing that they are "one ideology, a focused identity, [with] a global strategy," but we do not agree that *any* re-labeling would have this effect. On the contrary re-labeling it as a different kind of threat—as an organized criminal enterprise rather than a military struggle—can help resist terrorists' religious legitimation efforts,[15] better mobilize domestic Muslim support, and could revitalize non-Muslim domestic attitudes at a time when support for the war frame is waning. The focus on criminal activity is a natural one given the strong connection between criminal activity and terrorism.[16] Focusing on terrorism as a violation of international law accomplishes several purposes.

Framing Terrorism as a Violation of International Law Makes it Possible to Communicate Success

Americans understand that crime cannot be eliminated but that it can be contained. This means that when the coalition undertakes a successful operation it will be judged as to whether or not it decreased criminal activity, not whether or not it was a "total victory." Having attainable goals means that Americans are more likely to sustain support for the effort.

Framing Terrorism as a Violation of International Law Increases Awareness of the Problem as a Global One

This language reframes the discussion of terrorism against the backdrop of international standards. Hence, when the United States takes action against terrorism, we take action on behalf of the laws of the international community. This framing increases the opportunities for cooperation with other international actors who seek a more just world based on rule of law.

Framing Terrorism as a Violation of International Law Means that Terrorists are Reduced to the Level of Criminals

Positioning the conflict as a "war" confers a certain amount of legitimacy on terrorists, as a force defending a large collective interest. Reducing terrorists to criminals who seek money and power detracts from the allure of terrorism to possible recruits.

Framing Terrorism as a Violation of International Law Removes Any Connection of the Language to Religion

Referring to terrorists as criminals rather than "jihadis" completely separates them from the Muslim faith. Distancing the conflicts from religion means that we are more likely to win moderate Muslim cooperation because we will implicitly acknowledge the disconnection between terrorist activity and Islam. Conversely the current policy runs the constant risk of conflating "terrorist" with "Muslim." We cannot sustain this policy if we expect to win friends within the Muslim community.

The new language of terrorism will require development, refinement, and contributions from a variety of sources. However, within the same framework we utilized in the analysis of presidential speeches, we have devised the basis for a vocabulary of terrorism that distinguishes the current situation from traditional warfare (Table 2). From our newly established vocabulary, we have rewritten one of Bush's speeches from his campaign during the anniversary of 9/11 (see appendix). In our modification, we have left much of the President's words intact, but have changed the instances where the traditional war frame appears. We are confident that readers who give this a fair reading will find that the language equally as tough, determined, and unfavorable to the terrorists.

Table 2. New Characterizations of Terrorism

	US	THEM
Action	Policing, Building alliances, Increasing communications, Increasing independence	Steal, Lie, Cheat, Murder, Assassinate, Violate, Launder, Manipulate, Propagandize, Disrupt, Breach, Infringe, Transgress, Abuse, Assault
Goals	Decreased levels of activity, Removing key players, Maintaining order, Increasing cooperation, Law enforcement	Acquire power and money, Disrupt order, Break law, Bend international will, Control international actors
People	International peacekeepers	Enemies of international law, Criminals, Outlaws, Fugitives, Bandits, Assassins, Murders, Thieves, Thugs
Means	Freezing assets, Breaking up rackets/networks, Isolating activity, Containment, Imprisonment	Underground networks, Human rights abuses, Violence, Threats, Intimidation, Bombing, Recruiting
Environment	International order, Back Alleys, Dark Corners (Internet)	

Conclusion

With increasing doubts toward the Bush administration's policies in the War on Terror, new frameworks and vocabularies are necessary to revitalize public support. The War on Terror provides a new type of battle on an ambiguous front where victory will not be realized through peace treaties and flags atop hills. We have offered a new vocabulary, that of international crime, to reframe the War on Terror. This strategic new vocabulary paints terrorism as a persistent international problem and undermines the religious justification commonly asserted by Al Qaeda and other extremists. Removing the religious connotation of the War on Terror will help mobilize moderate Muslims in support of efforts that reduce international crime. Finally, reframing the War on Terror as an international crime problem allows international bodies to enforce existing law systems which already govern criminal acts.

Bush Speech Modified with New Vocabulary in Bold

President's Address to the Nation, September 11, 2006

> THE PRESIDENT: Good evening. Five years ago, this date—September the 11th—was seared into America's memory. Nineteen men attacked us with a barbarity unequaled in our history. They murdered people of all colors, creeds, and nationalities—and **violated international law and order**. Since that day, America **and members of the international community have enforced international norms and laws, policed these underground criminal networks, and hunted down the murderous outlaws**. Today, we are safer, but we are not yet safe. On this solemn night, I've asked for some of your time to discuss the nature of the threat still before us, what we are doing to protect our nation, and the building of a more hopeful Middle East that holds the key to peace for America and the world.
>
> On 9/11, our nation saw the face of evil. Yet on that awful day, we also witnessed something distinctly American: ordinary citizens rising to the occasion, and responding with extraordinary acts of courage. We saw courage in office workers who were trapped on the high floors of burning skyscrapers—and called home so that their last words to their families would be of comfort and love. We saw courage in passengers aboard Flight 93, who recited the 23rd Psalm—and then charged the cockpit. And we saw courage in the Pentagon staff who made it out of the flames and smoke—and ran back in to answer cries for help. On this day, we remember the innocent who lost their

lives—and we pay tribute to those who gave their lives so that others might live.

For many of our citizens, the wounds of that morning are still fresh. I've met firefighters and police officers who choke up at the memory of fallen comrades. I've stood with families gathered on a grassy field in Pennsylvania, who take bittersweet pride in loved ones who refused to be victims. I've sat beside young mothers with children who are now five years old—and still long for the daddies who will never cradle them in their arms. Out of this suffering, we resolve to honor every man and woman lost. And we seek their lasting memorial in a safer and more hopeful world.

Since the horror of 9/11, we've learned a great deal about **these criminals**. We have learned that they are evil and kill without mercy—but not without purpose. We have learned that they form a global **criminal** network **driven by** a **violent self-interest** that **disregards international order**, rejects tolerance, and despises all dissent. And we have learned that their goal is to **disrupt international norms to the point** where women are prisoners in their homes, men are beaten for missing prayer meetings, and terrorists have a safe haven to plan and launch **criminal** attacks on America and other civilized nations. **The hunt for these criminals has changed the international landscape and the world is united against these murderous thugs.**

Our nation is being tested in a way that we have not been since the start of the Cold War. We saw what a handful of our enemies can do with box-cutters and plane tickets. We hear their threats to launch even more terrible attacks on our people. And we know that if they were able to get their hands on weapons of mass destruction, they would use them against us. We face an enemy determined to bring death and suffering into our homes. America did not ask for this **violence**, and every American wishes it were over. So do I. But the **hunt** is not over—and it will not be over until we **capture every fugitive**. If we do not **contain** these enemies now, we will leave our children to face a Middle East overrun by terrorist states and radical dictators armed with nuclear weapons. We are in a **struggle** that will set the course for this new century—and determine the destiny of millions across the world.

For America, 9/11 was more than a tragedy—it changed the way we look at the world. On September the 11th, we resolved that we would go on the **hunt for** our enemies, and we would not distinguish between the terrorists and those who harbor or support them. So we helped drive the Taliban from power in Afghanistan. We put al Qaeda on the run, and killed or captured most of those who planned the 9/11 attacks, including the man believed to be the mastermind, Khalid Sheik Mohammed. He and other suspected terrorists have been questioned by the Central Intelligence Agency, and they provided valuable information that has helped stop **criminal** attacks in America and across the world. Now these men have been transferred to Guantanamo Bay, so they can be held to account for their actions. Osama bin Laden and other **criminal agents** are still in hiding. Our message to them is clear: No matter how long it takes, America will find you, and we will bring you to justice.

On September the 11th, we learned that America must confront threats before they reach our shores, whether those threats come from terrorist networks. I'm often asked why we're in Iraq when Saddam Hussein was not responsible for the 9/11 attacks. The answer is that the regime of Saddam Hussein was a clear threat. My administration, the Congress, and the United Nations saw the threat—and after 9/11, Saddam's regime posed a risk that the world could not afford to take. The world is safer because Saddam Hussein is no longer in power. And now the challenge is to help the Iraqi people build a democracy that fulfills the dreams of the nearly 12 million Iraqis who came out to vote in free elections last December.

Al Qaeda and other **criminal** extremists from across the world have come to Iraq to stop the rise of a free society in the heart of the Middle East. They have joined the remnants of Saddam's regime and other armed groups to foment sectarian violence and drive us out. Our enemies in Iraq are tough and they are committed—but so are Iraqi and coalition forces. We're adapting to stay ahead of the enemy, and we are carrying out a clear plan to ensure that a democratic Iraq succeeds.

We're training Iraqi troops so they can defend their nation. We're helping Iraq's unity government grow in strength and serve its people. We will not leave until this work is done. Whatever mistakes have been made in Iraq, the worst mistake would be to think that if we pulled out, the terrorists would leave us alone. They will not leave us alone. They will follow us. The safety of America depends on the outcome of the battle in the streets of Baghdad. Osama bin Laden calls this fight "the Third World War"—and he says that victory for the terrorists in Iraq will mean America's "defeat and disgrace forever." If we yield Iraq to men like bin Laden, our enemies will be emboldened; they will gain a new safe haven; they will use Iraq's resources to fuel their extremist movement. We will not allow this to happen. America will stay in the fight. Iraq will be a free nation, and **an international actor in terrorist crime reduction**.

We can be confident that our coalition will succeed because the Iraqi people have been steadfast in the face of unspeakable violence. And we can be confident in victory because of the skill and resolve of America's Armed Forces. Every one of our troops is a volunteer, and since the attacks of September the 11th, more than 1.6 million Americans have stepped forward to put on our nation's uniform. In Iraq **and** Afghanistan, the men and women of our military are making great sacrifices to keep us safe. Some have suffered terrible injuries—and nearly 3,000 have given their lives. America cherishes their memory. We pray for their families. And we will never back down from the work they have begun.

We also honor those who toil day and night to keep our homeland safe, and we are giving them the tools they need to protect our people. We've created the Department of Homeland Security. We have torn down the wall that kept law enforcement and intelligence from sharing information. We've tightened security at our airports and seaports and borders, and we've created new programs to monitor enemy bank records and phone calls. Thanks to the hard work of our law enforcement and intelligence professionals, we have broken up terrorist cells in our midst and saved American lives.

Five years after 9/11, our enemies have not succeeded in launching another attack on our soil, but they've not been idle. Al Qaeda and those inspired by its **criminal** ideology have carried out terrorist attacks in more than two dozen nations. And just last month, they were foiled in a plot to blow up passenger planes headed for the United States. They remain determined to attack America and kill our citizens—and we are determined to stop them. We'll continue to give the men and women who protect us every resource and legal authority they need to do their jobs.

In the first days after the 9/11 attacks I promised to use every element of national power to fight the terrorists, wherever we find them. One of the **most effective tools** in our **possession** is the power of freedom. The terrorists fear freedom as much as they do our firepower. They are thrown into panic at the sight of an old man pulling the election lever, girls enrolling in schools, or families worshiping God in their own traditions. They know that given a choice, people will choose freedom over their extremist ideology. So their answer is to deny people this choice by raging against the forces of freedom and moderation. We are fighting to maintain the way of life enjoyed by free nations. And we're fighting for the possibility that good and decent people across the Middle East can raise up societies based on freedom and tolerance and personal dignity.

We are now in the early hours of this struggle between **order and disorder**. Amid the violence, some question whether the people of the Middle East want their freedom, and whether the forces of moderation can prevail. For 60 years, these doubts guided our policies in the Middle East. And then, on a bright September morning, it became clear that the calm we saw in the Middle East was only a mirage. Years of pursuing stability to promote peace had left us with neither. So we changed our policies, and committed America's influence in the world to advancing freedom and democracy as the great alternatives to repression and radicalism.

With our help, the people of the Middle East are now stepping forward to claim their freedom. From Kabul to Baghdad to Beirut, there are brave men and women risking their lives each day for the same freedoms that we enjoy. And they have one question for us: Do we have the confidence to do in the Middle East what our fathers and grandfathers accomplished in Europe and Asia? By standing with democratic leaders and reformers, by giving voice to the hopes of decent men and women, we're offering a path away from radicalism. And we are enlisting the most powerful force for peace and moderation in the Middle East: the desire of millions to be free.

At the start of this young century, America looks to the day when the people of the Middle East leave the desert of despotism for the fertile gardens of liberty, and resume their rightful place in a world of peace and prosperity. We look to the day when the nations of that region recognize their greatest resource is not the oil in the ground, but the talent and creativity of their people. We look to the day when moms and dads throughout the Middle East see a future of hope and opportunity for their children. And when that good day comes, the clouds of **violence** will part, the appeal of radicalism will decline, and we will leave our children with a better and safer world.

On this solemn anniversary, we rededicate ourselves to this cause. Our nation has endured trials, and we face a difficult road ahead. **Containing terrorism** will require the determined efforts of a unified **international community**, and we must put aside our differences and work together. We will **break up criminal networks and bring our enemies to justice**. We will protect our people. And we will lead the 21st century into a shining age of human liberty.

Earlier this year, I traveled to the United States Military Academy. I was there to deliver the commencement address to the first class to arrive at West Point after the **crimes** of September the 11th. That day I met a proud mom named RoseEllen Dowdell. She was there to watch her son, Patrick, accept his commission in the finest Army the world has ever known. A few weeks earlier, RoseEllen had watched her other son, James, graduate from the Fire Academy in New York City. On both these days, her thoughts turned to someone who was not there to share the moment: her husband, Kevin Dowdell. Kevin was one of the 343 firefighters who rushed to the burning towers of the World Trade Center on September the 11th—and never came home. His sons lost their father that day, but not the passion for service he instilled in them. Here is what RoseEllen says about her boys: "As a mother, I cross my fingers and pray all the time for their safety—but as worried as I am, I'm also proud, and I know their dad would be, too."

Our nation is blessed to have young Americans like these—and we will need them. Dangerous enemies have declared their intention to destroy our way of life. They're not the first to try, and their fate will be the same as those who tried before. Nine-Eleven showed us why. The **crimes** were meant to bring us to our knees, and they did, but not in the way the terrorists intended. Americans united in prayer, came to the aid of neighbors in need, and resolved that our enemies would not have the last word. The spirit of our people is the source of America's strength. And we go forward with trust in that spirit, confidence in our purpose, and faith in a loving God who made us to be free.

Thank you, and may God bless you.

Original Bush Speech with War Framing

President's Address to the Nation, September 11, 2006

THE PRESIDENT: Good evening. Five years ago, this date—September the

11th—was seared into America's memory. Nineteen men attacked us with a barbarity unequaled in our history. They murdered people of all colors, creeds, and nationalities—and made war upon the entire free world. Since that day, America and her allies have taken the offensive in a war unlike any we have fought before. Today, we are safer, but we are not yet safe. On this solemn night, I've asked for some of your time to discuss the nature of the threat still before us, what we are doing to protect our nation, and the building of a more hopeful Middle East that holds the key to peace for America and the world.

On 9/11, our nation saw the face of evil. Yet on that awful day, we also witnessed something distinctly American: ordinary citizens rising to the occasion, and responding with extraordinary acts of courage. We saw courage in office workers who were trapped on the high floors of burning skyscrapers—and called home so that their last words to their families would be of comfort and love. We saw courage in passengers aboard Flight 93, who recited the 23rd Psalm—and then charged the cockpit. And we saw courage in the Pentagon staff who made it out of the flames and smoke—and ran back in to answer cries for help. On this day, we remember the innocent who lost their lives—and we pay tribute to those who gave their lives so that others might live.

For many of our citizens, the wounds of that morning are still fresh. I've met firefighters and police officers who choke up at the memory of fallen comrades. I've stood with families gathered on a grassy field in Pennsylvania, who take bittersweet pride in loved ones who refused to be victims—and gave America our first victory in the war on terror. I've sat beside young mothers with children who are now five years old—and still long for the daddies who will never cradle them in their arms. Out of this suffering, we resolve to honor every man and woman lost. And we seek their lasting memorial in a safer and more hopeful world.

Since the horror of 9/11, we've learned a great deal about the enemy. We have learned that they are evil and kill without mercy—but not without purpose. We have learned that they form a global network of extremists who are driven by a perverted vision of Islam—a totalitarian ideology that hates freedom, rejects tolerance, and despises all dissent. And we have learned that their goal is to build a radical Islamic empire where women are prisoners in their homes, men are beaten for missing prayer meetings, and terrorists have a safe haven to plan and launch attacks on America and other civilized nations. The war against this enemy is more than a military conflict. It is the decisive ideological struggle of the 21st century, and the calling of our generation.

Our nation is being tested in a way that we have not been since the start of the Cold War. We saw what a handful of our enemies can do with box-cutters and plane tickets. We hear their threats to launch even more terrible attacks on our people. And we know that if they were able to get their hands on weapons of mass destruction, they would use them against us. We face an enemy determined to bring death and suffering into our homes. America did not ask for this war, and every American wishes it were over. So do I. But the war is not over—and it will not be over until either we or the extremists emerge victorious. If we do not defeat these enemies now, we will leave our children to face a Middle East overrun by terrorist states and radical dictators armed with nuclear weapons. We are in a war that will set the course for this new century—and determine the destiny of millions across the world.

For America, 9/11 was more than a tragedy—it changed the way we look at the world. On September the 11th, we resolved that we would go on the offense against our enemies, and we would not distinguish between the terrorists and those who harbor or support them. So we helped drive the Taliban from power in Afghanistan. We put al Qaeda on the run, and killed or captured most of those who planned the 9/11 attacks, including the man believed to be the mastermind, Khalid Sheik Mohammed. He and other suspected terrorists have been questioned by the Central Intelligence Agency, and they provided valuable information that has helped stop attacks in America and across the world. Now these men have been transferred to Guantanamo Bay, so they can be held to account for their actions. Osama bin Laden and other terrorists are still in hiding. Our message to them is clear: No matter how long it takes, America will find you, and we will bring you to justice.

On September the 11th, we learned that America must confront threats before they reach our shores, whether those threats come from terrorist networks or terrorist states. I'm often asked why we're in Iraq when Saddam Hussein was not responsible for the 9/11 attacks. The answer is that the regime of Saddam Hussein was a clear threat. My administration, the Congress, and the United Nations saw the threat—and after 9/11, Saddam's regime posed a risk that the world could not afford to take. The world is safer because Saddam Hussein is no longer in power. And now the challenge is to help the Iraqi people build a democracy that fulfills the dreams of the nearly 12 million Iraqis who came out to vote in free elections last December.

Al Qaeda and other extremists from across the world have come to Iraq to stop the rise of a free society in the heart of the Middle East. They have joined the remnants of Saddam's regime and other armed groups to foment sectarian violence and drive us out. Our enemies in Iraq are tough and they are committed—but so are Iraqi and coalition forces. We're adapting to stay ahead of the enemy, and we are carrying out a clear plan to ensure that a democratic Iraq succeeds.

We're training Iraqi troops so they can defend their nation. We're helping Iraq's unity government grow in strength and serve its people. We will not leave until this work is done. Whatever mistakes have been made in Iraq, the worst mistake would be to think that if we pulled out, the terrorists would leave us alone. They will not leave us alone. They will follow us. The safety of America depends on the outcome of the battle in the streets of Baghdad. Osama bin Laden calls this fight "the Third World War"—and he says that victory for the terrorists in Iraq will mean America's "defeat and disgrace forever." If we yield Iraq to men like bin Laden, our enemies will be emboldened; they will gain a new safe haven; they will use Iraq's resources to fuel their extremist movement. We will not allow this to happen. America will stay in the fight. Iraq will be a free nation, and a strong ally in the war on terror.

We can be confident that our coalition will succeed because the Iraqi people have been steadfast in the face of unspeakable violence. And we can be confident in victory because of the skill and resolve of America's Armed Forces. Every one of our troops is a volunteer, and since the attacks of September the 11th, more than 1.6 million Americans have stepped forward to put on our nation's uniform. In Iraq, Afghanistan, and other fronts in the war on terror, the men and women of our military are making great sacrifices to keep us safe. Some have suffered terrible injuries—and nearly 3,000 have given their lives. America cherishes their memory. We pray for their families. And we will never back down from the work they have begun.

We also honor those who toil day and night to keep our homeland safe, and we are giving them the tools they need to protect our people. We've created the Department of Homeland Security. We have torn down the wall that kept law enforcement and intelligence from sharing information. We've tightened security at our airports and seaports and borders, and we've created new programs to monitor enemy bank records and phone calls. Thanks to the hard work of our law enforcement and intelligence professionals, we have broken up terrorist cells in our midst and saved American lives.

Five years after 9/11, our enemies have not succeeded in launching another attack on our soil, but they've not been idle. Al Qaeda and those inspired by its hateful ideology have carried out terrorist attacks in more than two dozen nations. And just last month, they were foiled in a plot to blow up passenger planes headed for the United States. They remain determined to attack America and kill our citizens—and we are determined to stop them. We'll continue to give the men and women who protect us every resource and legal authority they need to do their jobs.

In the first days after the 9/11 attacks I promised to use every element of national power to fight the terrorists, wherever we find them. One of the strongest weapons in our arsenal is the power of freedom. The terrorists fear freedom as much as they do our firepower. They are thrown into panic at the sight of an old man pulling the election lever, girls enrolling in schools, or families worshiping God in their own traditions. They know that given a choice, people will choose freedom over their extremist ideology. So their answer is to deny people this choice by raging against the forces of freedom and moderation. This struggle has been called a clash of civilizations. In truth, it is a struggle for civilization. We are fighting to maintain the way of life enjoyed by free nations. And we're fighting for the possibility that good and decent people across the Middle East can raise up societies based on freedom and tolerance and personal dignity.

We are now in the early hours of this struggle between tyranny and freedom. Amid the violence, some question whether the people of the Middle East want their freedom, and whether the forces of moderation can prevail. For 60 years, these doubts guided our policies in the Middle East. And then, on a bright September morning, it became clear that the calm we saw in the Middle East was only a mirage. Years of pursuing stability to promote peace had left us with neither. So we changed our policies, and committed America's influence in the world to advancing freedom and democracy as the great alternatives to repression and radicalism.

With our help, the people of the Middle East are now stepping forward to claim their freedom. From Kabul to Baghdad to Beirut, there are brave men and women risking their lives each day for the same freedoms that we enjoy. And they have one question for us: Do we have the confidence to do in the Middle East what our fathers and grandfathers accomplished in Europe and Asia? By standing with democratic leaders and reformers, by giving voice to the hopes of decent men and women, we're offering a path away from radicalism. And we are enlisting the most powerful force for peace and moderation in the Middle East: the desire of millions to be free.

Across the broader Middle East, the extremists are fighting to prevent such a future. Yet America has confronted evil before, and we have defeated it—sometimes at the cost of thousands of good men in a single battle. When Franklin Roosevelt vowed to defeat two enemies across two oceans, he could not have foreseen D-Day and Iwo Jima—but he would not have been surprised at the outcome. When Harry Truman promised American support for free peoples resisting Soviet aggression, he could not have foreseen the rise of the Berlin Wall—but he would not have been surprised to see it brought down. Throughout our history, America has seen liberty challenged, and every time, we have seen liberty triumph with sacrifice and determination.

At the start of this young century, America looks to the day when the people of the Middle East leave the desert of despotism for the fertile gardens of liberty, and resume their rightful place in a world of peace and prosperity. We look to the day when the nations of that region recognize their greatest resource is not the oil in the ground, but the talent and creativity of their people. We look to the day when moms and dads throughout the Middle East see a future of hope and opportunity for their children. And when that good day comes, the clouds of war will part, the appeal of radicalism will decline, and we will leave our children with a better and safer world.

On this solemn anniversary, we rededicate ourselves to this cause. Our nation has endured trials, and we face a difficult road ahead. Winning this war will require the determined efforts of a unified country, and we must put aside our differences and work together to meet the test that history has given us. We will defeat our enemies. We will protect our people. And we will lead the 21st century into a shining age of human liberty.

Earlier this year, I traveled to the United States Military Academy. I was there to deliver the commencement address to the first class to arrive at West Point after the attacks of September the 11th. That day I met a proud mom named RoseEllen Dowdell. She was there to watch her son, Patrick, accept his commission in the finest Army the world has ever known. A few weeks earlier, RoseEllen had watched her other son, James, graduate from the

Fire Academy in New York City. On both these days, her thoughts turned to someone who was not there to share the moment: her husband, Kevin Dowdell. Kevin was one of the 343 firefighters who rushed to the burning towers of the World Trade Center on September the 11th—and never came home. His sons lost their father that day, but not the passion for service he instilled in them. Here is what RoseEllen says about her boys: "As a mother, I cross my fingers and pray all the time for their safety—but as worried as I am, I'm also proud, and I know their dad would be, too."

Our nation is blessed to have young Americans like these—and we will need them. Dangerous enemies have declared their intention to destroy our way of life. They're not the first to try, and their fate will be the same as those who tried before. Nine-Eleven showed us why. The attacks were meant to bring us to our knees, and they did, but not in the way the terrorists intended. Americans united in prayer, came to the aid of neighbors in need, and resolved that our enemies would not have the last word. The spirit of our people is the source of America's strength. And we go forward with trust in that spirit, confidence in our purpose, and faith in a loving God who made us to be free. Thank you, and may God bless you.

Notes

1. *War on terror update: An upswing in confidence.* (2006, October 28). Rasmussen.
2. *National Security Strategy of the United States of America.* (2002, Sept). *The White House*, p. 1.
3. *Permanent Select Committee on Intelligence.* (2006, Sept 20). Al-Qaeda: The many faces of an Islamist extremist threat, p. 6.
4. Reynolds, P. (2007, April 17). Declining use of 'war on terror.' *BBC News.*
5. Perlez, J. (2007, April 17). Briton criticizes U.S.'s use of 'War on Terror.' *New York Times.*
6. Gilbert, C. (2006, September 12). Sept. 11: Five years later Feingold decries creation of term 'Islamic fascism'. *Milwaukee Journal Sentinel*, p. 10.
7. Robinson, M. (2006, Winter) Law, language, and principle in the "war on terror." *Human rights magazine.*
8. Medhurst, M. J. (1990). Rhetoric and cold war: A strategic approach. In Medhurst, M. J., Ivie, R. L., Wander, P., and Scott, R. L. (Eds.) *Cold war rhetoric: Strategy, metaphor, and ideology.* New York: Greenwood Press, p. 19.
9. Ivie, R. L. (1974). Presidential motives for war. *Quarterly Journal of Speech, 60*, p. 344.
10. Bush, G. W. (2006, Aug 31). *President Bush addresses American Legion National Convention. The White House*; Bush, G. W. (2006, Sept 5). *President discusses Global War on Terror. The White House*; Bush, G. W. (2006, Sept 7). *President Bush discusses*

progress in the Global War on Terror. The White House; Bush, G. W. (2006, Sept 11). *President's address to the nation. The White House.* Bush, G. W. (2006, Sept 29). *President Bush discusses Global War on Terror. The White House.*

11. Baker, P. & VandeHei, J. (2006, August 31). Bush team casts foes as defeatist; Blunt rhetoric signals a new thrust. *Washington Post.* Retrieved November 2, 2006 from Lexis-Nexis.com.
12. Justus, Z. S. & Hess, A. Chapter 6, this volume.
13. Justus, Z. S. & Hess, A. Chapter 6, this volume.
14. Phares, W. (2007, April). *British Minister fails the War of Ideas.* Counterterrorism Blog.
15. See Corman, S.R. and Schiefelbein, J.S. (2006). *Communication and media strategy in the jihadi war of ideas.* Report #0601, Consortium for Strategic Communication, Arizona State University.
16. Gartenstein-Ross, D. & Dabruzzi, K. (2007). *The convergence of crime and terror: Law enforcement opportunities and perils. The Center for Policing Terrorism.*

PART III

Crafting a New Communication Policy

CHAPTER EIGHT

A New Communication Model for the 21st Century

From Simplistic Influence to Pragmatic Complexity

STEVEN R. CORMAN, ANGELA TRETHEWEY, AND H. L. GOODALL, JR.
WITH CONTRIBUTIONS FROM ROBERT D. MCPHEE

Introduction

In the global war of ideas, the United States finds itself facing a systems problem that cannot be solved by simply delivering the right message. The question is not "how can we construct a more persuasive message?" Rather it is "what kind of reality has this particular system [that we are trying to influence] constructed for itself?"

The present strategic communication efforts by the U.S. and its allies rest on an outdated, 20th-century *message influence model* that is no longer effective in the complex global war of ideas. Relying on this model, our well-intentioned communication has become dysfunctional. Rather than drawing the world into a consensus on issues of terrorism, diplomacy, and international security, it instead unwittingly contributes to our diminished status among world opinion leaders and furthers the recruitment goals of violent extremists.

In this chapter we explain why message influence strategies fail and what must be done to break the cycle of communication dysfunction. Changing communication systems requires, first, understanding the dynamics at work; and, second, using communication as a strategy to disrupt and perturb existing systems such that they can begin to organize around new meaning-making frameworks. After describing a new *pragmatic complexity model*, we offer four principles of effective communication in the global war of idcas based on this model: (1) Deemphasize control and embrace complexity, (2) replace repetition with variation, (3) consider disruptive moves, and (4) expect and plan for failure.

Communication is a vital tool of terrorist groups. Violent extremists use communication to spread their ideology, legitimize their actions, recruit new supporters, and intimidate enemies.[1] They do these things using overt messages sent via personal interaction, mass media, and Internet postings, as well as through secondary coverage of their violent activities through similar channels.

The United States and its allies in the West have a strikingly similar set of communication goals. They seek to spread a counter-ideology of Western values like democracy, legitimize their actions, gain public support, and intimidate the terrorists and their supporters. Public communication is therefore of special strategic importance in the Global War on Terrorism (GWOT).

Message Influence Model

The communication model underlying current Western strategic communication practices dates back at least to the 1950s. In 1960, Berlo published a text[3] describing the perspective, which he developed to support a series of workshops conducted for the U.S. International Cooperation Administration.[4] It draws heavily on an analogy comparing human communication to transmission of messages over a telephone system. Shannon and Weaver[5] originated this idea, defining communication as a process in which one mind uses messages to affect another mind. Their model (Figure 1) assumes that there is an information source that has a message encoded by a transmitter. The transmitter converts the message into a signal which is sent through some channel, during which time it may be degraded to some extent by noise. The signal enters a receiver, which decodes it back into a message, which arrives at the destination.

Figure 1. Shannon and Weaver's Model

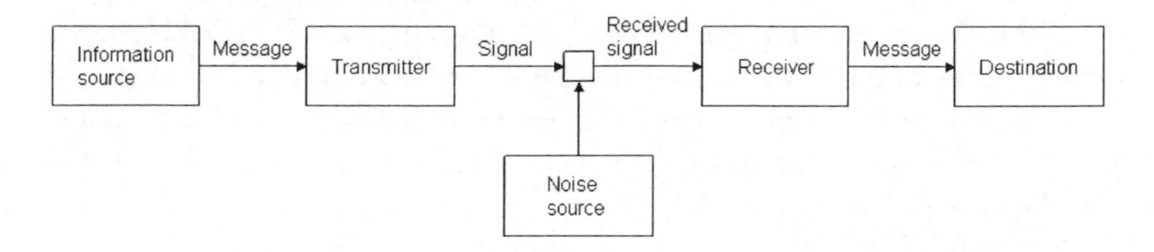

Berlo applies this model straightforwardly to human communication. He begins with a *source* that has "ideas, needs, intentions, information, and a purpose for communicating." These things are formulated as a *message* which is translated into a systematic set of symbols by an *encoder* that employs the motor skills of the communicator. The encoded message is sent via some *channel* (a particular medium of communication) to the *receiver*, who uses a *decoder* to "retranslate" the symbols into a usable form.[6] We call this a *message influence model* because it conceptualizes messages as a vehicle for carrying information from a source to a receiver. The purpose of the message is to influence the receiver to understand the information in the same way as the source, if not persuade him or her to change attitudes or act in a particular way.

One of the implications of this view is that failures are a matter of interference of one kind or another with the transmission of the message. Berlo describes this as the *fidelity* of the message,[7] which determines the message's effect. One source of infidelity is noise occurring in the channel. It can usually be tolerated (for example, we can successfully talk even on a noisy phone connection), or overcome through the repetition of the same message, or even avoided altogether by choosing a better channel. Outright distortion of messages occurs in the encoding or decoding stages. Distortion occurs because communicators lack sufficient skill to faithfully translate the information to or from symbols, or their culture or individual attitudes corrupt the translation process in some way.

A key underlying assumption of the message influence model is that communication will be successful unless the factors just described interfere with the sender/receiver connection. Accordingly "best practices" can be employed by influence-seeking sources to promote fidelity in their transmissions. Simple, concise messages are

The old model assumes that communication will automatically be successful unless there is a bad connection.

superior to complicated ones because they are easier to encode and decode faithfully. Messages can be repeated to insure that unskilled receivers have multiple chances to get it right, making the transmission more reliable despite the presence of noise. The sender can also try to understand the attitudes and cultural context of the receiver and then use his or her skill to encode messages that are least likely to be distorted by them. Table 1 summarizes characteristics of the message influence model.

Table 1. Message Influence Model Summary

Communication concept	Sending messages and "signals" to a well-defined audience
Constraint	Communicator skill
Principles	Insure message fidelity Influence attitudes/beliefs/behaviors Avoiding misunderstanding
Expectation	Success

The message influence model is not some obscure concept that occurred in one text published in 1960. It is a sign of the thinking of the times. The 1957 Vance Packard classic *The Hidden Persuaders* notes that only in that decade did social science methods start to influence the ideas of political communication. The then-chairman of the Republican Party believed that political communication works in the same way as successful advertising: "You sell your candidates and your programs the way a business sells its products."[8] Other texts and scholarly works on communication expressed similar ideas, and (with some minor modifications) the message influence model continues to inform strategic communication practice to this day. For example in the contemporary *Public Relations Handbook*,[9] Fawkes describes a modification of the model that has messages going not only from source to receiver but from receiver back to source, in a circular fashion. A further elaboration is the Westley-McLean model (Figure 2). Here communicator A gathers information from the environment (Xs) and formulates a message (X') that moves through the channel or gatekeeper (C) who may change it (X") before it gets to the public (B) who

Figure 2. The Westley-McLean Model

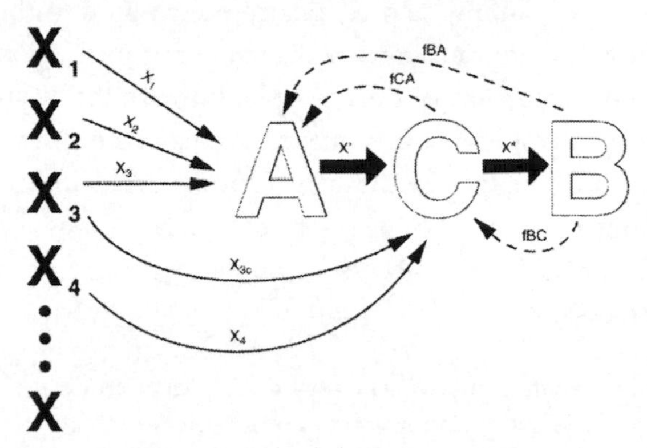

provides feedback (f) to the various stages. Fawkes acknowledges that interpretive processes of the audience and characteristics of the media can have important effects. Yet it is clear that despite the addition of communicators, feedback paths, interpretive processes, and so on, the message influence model is alive and well in the public relations discipline, and it still conceives communication in the same linear terms described above.

The old model continues to inform strategic communication practice to this day.

The message influence model also pervades post-9/11 thinking about public diplomacy, public affairs, information operations, and media strategy in the United States government. In January of 2003 President Bush signed an Executive Order creating the White House Office of Global Communications.[10] Its stated mission is to "ensure consistency in messages that will promote the interests of the United States abroad, prevent misunderstanding, build support for and among coalition partners of the United States, and inform international audiences." It would do this, in part, by establishing "information teams" that would "disseminate accurate and timely information about topics of interest to the on-site news media."

In its assessment published the next year, the 9/11 Commission concluded that failure to adhere to the message influence model was part of the problem: "The U.S. Government must define what its message is, what it stands for" and it "must do more to communicate its message."[11] For this purpose it suggested new initiatives in television and radio broadcasting in strategically important areas. Shortly thereafter, a GAO investigation complained of a lack of interagency communication strategy, concluding that "the absence of such a strategy complicates the task of conveying consistent messages to overseas audiences."[12]

The old model pervades post-9/11 thinking about public diplomacy, public affairs, information operations, and media strategy.

Those who advise the government focus on message influence too. For example, Newton

Minow complained in a 2003 memorandum that "we have failed to use the power of ideas" and we should be "explaining and advocating our values to the world." We could do this by broadcasting messages that "make our ideas clear not just to leaders in the Muslim world, but to those people in the street." Our superior skill at delivering messages would insure success: "We have the smartest, most talented, and most creative people in the world in our communications industries" who "will volunteer eagerly to help get our message across."[13] A Rand paper in 2004 also concluded that success in public diplomacy is a matter of delivering the right message:

> As important as it is to communicate America's history of support and defense of Muslim populations, it is equally important to communicate the rationale motivating these policies. In these instances, U.S. policies reflected and furthered the values of democracy, tolerance, the rule of law, and pluralism. The overarching message public diplomacy should convey is that the United States tries, although it does not always succeed, to further these values regardless of the religion, ethnicity, or other characteristics of the individuals and groups involved. Highlighting the instances in which the United States has benefited Muslim populations by acting on these values may make this point more salient.[14]

Why Message Influence Fails

Despite its pervasiveness and "taken-for-granted" appeal, the message influence model has failed to deliver success. This is not for lack of well-intended efforts by government spokespersons to enhance the U.S. global image. Instead, the message influence model—used by our diplomats, government officials, and key communicators—is flawed because it fails to respond to the complexities of communication as a *meaning-making* process.

The message influence model assumes, incorrectly, that communication is the transfer of "meanings from person to person"[15] and that the message sent is the one that counts. The problem is that *a meaning cannot simply be transferred*, like a letter mailed from point A to point B. Instead, listeners create meanings from messages based on factors like autobiography, history, local context, culture, language/symbol systems, power relations, and immediate personal needs. We should assume that meanings listeners create in their minds will probably not be identical to those intended by the sender. As several decades of communication research have shown, the message *received* is the one that really counts.[16]

These shortcomings of the message influence model were evident when Karen Hughes embarked on her "listening tour" of the Middle East in 2005. She hoped to begin the process of improving the damaged image of the U.S. by inspiring audiences with a vision of American democracy. Pursuing President Bush's strategy of delivering a clear and consistent message, Ms. Hughes said to a group of 500 Saudi women:

A meaning cannot simply be transferred, like a letter mailed from point A to point B.

> I feel, as an American woman, that my ability to drive is an important part of my freedom. It has allowed me to work during my career. It has allowed me to go to the grocery store and shop for my family. It allows me to go to the doctor.[17]

The intent of this message was to highlight the freedoms that accrue to American women, particularly in comparison to the audience's government, which (among other restrictions) bans women from driving. Yet, the message fell flat. Not only did Hughes fail to persuade the Saudi women, she inadvertently offended her audience. In response to Hughes, the Egyptian daily *Al-Jumhuriya* responded, "We in Egypt, and everywhere else, don't need America's public relations campaign."[18] Clearly, in this case just delivering a clear message, or repeating it from country to country, was not enough.

Another problem with traditional thinking about communication is that messages are always interpreted within a larger, ongoing communication system. In the language of communication science, communication is the medium through which individuals and groups construct their social realities. Once a system—a social reality—is created, it has a tendency to sustain itself even in the face of contradictory information and persuasive campaigns. Members of the system, routinely and often unconsciously, work to preserve the existing framework of meaning. To accomplish this they interpret messages in ways that "fit" the existing scheme, rather than in ways that senders may intend. There is no "magic bullet"—no single message, however well crafted—that can be delivered within the existing system that is likely to change it.

This dynamic is illustrated by the efforts of the United States to promote democracy in the Middle East, which has been a staple of U.S. foreign policy under the Bush administration. In a November 2003 speech, the President said:

> The establishment of a free Iraq at the heart of the Middle East will be a watershed event in the global democratic revolution. Sixty years of Western

> nations excusing and accommodating the lack of freedom in the Middle East did nothing to make us safe—because in the long run, stability cannot be purchased at the expense of liberty. As long as the Middle East remains a place where freedom does not flourish, it will remain a place of stagnation, resentment, and violence ready for export. And with the spread of weapons that can bring catastrophic harm to our country and to our friends, it would be reckless to accept the status quo. Therefore, the United States has adopted a new policy, a forward strategy of freedom in the Middle East.[19]

While it is hard for anyone in the West to imagine disagreeing with goals of freedom and democracy, Salafi extremists interpreted this message as yet another attempt by the Western Crusaders to impose their foreign values on Muslims. It was met with a fatwa from Abu Muhammad al-Maqdisi, the "key contemporary ideologue in the Jihadi intellectual universe,"[20] declaring democracy a "religion" that is at odds with Islamic principles of monotheism. As a result, the more the United States promotes its goal of democracy for Muslims, the more evidence the extremists have to reproduce their Crusader analogy.[21]

Repetition of a message can amplify meaning-making problems and damage the sender's credibility.

This self-preservation property of communication systems means that strategies highlighting the repetition of a seemingly clear and straightforward message can instead amplify meaning-making problems and damage the sender's credibility. In Western cultures, we rely on "talking cures" to creatively solve our problems in mutually beneficial ways. But if both parties are not in interpretive alignment, this technique can breed more problems than it solves.

Gregory Bateson describes a "mutually aggravating spiral by which each person's response to the other's behavior provokes more exaggerated forms of the divergent behavior."[22] The pattern commonly appears in relationships between husbands and wives. A woman asks her husband to call if he is going to be late from work. He interprets this as controlling his behavior, so he pulls away. Not only does he not call, but he tells her that he "needs some space," and perhaps plans to spend a weekend with his friends. This makes her fear that their relationship is in trouble, which encourages her to ask for even more connection, and so on. This spiraling pattern is common in families, between adults and children, and in interactions between people of different cultures.

The very same pattern is at work on the global landscape. The ongoing "veil affair" in France is an example. The French government has banned "ostentatious" religious symbols (including the head coverings worn by some Muslim women). Though it was officially intended to preserve religious tolerance and the spirit of a secular state (laïcité), the ban was interpreted by many Muslims as discriminatory, an affront to religious freedom, and evidence of French racism toward Muslims. Muslim women who resisted the ban by wearing veils in public were interpreted, at best, as defiant and, at worst, as an affront to French cultural values. The dynamics of this conflict are more complex than we can chart here, but it seems clear that both parties have adopted a failed message influence approach to change.

In the global war of ideas, the United States finds itself facing a systems problem too, one that cannot be solved by simply delivering the "right message." The question is not "how can we construct a more persuasive message?" Rather it is "what kind of reality has this particular system [that we are trying to influence] constructed for itself?"[23] Breaking dysfunctional communication systems requires, first, understanding the systems dynamics at work; and, second, using communication as a strategy to disrupt and perturb existing systems such that they can begin to organize around new meaning-making frameworks. Next we propose a new communication strategy for achieving this important goal.

Pragmatic Complexity Model

The shortcomings of the message influence model just described make it clear that we need an updated way of thinking about strategic communication. This is not to say that we can go without messages, or that it would be good to have unclear, inconsistent messages sent by unskilled communicators. Instead we call for an updated view of the *process* surrounding the communication of messages that avoids the simplistic view of the old message influence model, provides more realistic expectations about their impact, offers a new set of communication strategies, and in the long run leads to more strategic success.

Communication is a complex process of interpreting one another's actions and making attributions about thoughts, motivations, and intentions.

The new model we propose, pragmatic complexity (PCOM), draws ideas from the so-called "new systems" perspectives, especially the communication theory of Niklas Luhmann.[24] For him, communication is not an act of one mind transmitting a message to another mind. It is a property of a complex system in which participants interpret one another's actions and make attributions about the thoughts, motivations, intentions, etc., behind them. The issuing of a message by one party and its receipt by another may initiate this process, but that is far from the end of the story.

The system is complex because of a *double contingency* that involves the participants. In the simplest case of a communication system with two participants A and B, we can describe this constraint as follows:

- The success of A's behavior depends not only on external conditions, but on what B does and thinks.
- What B does and thinks is influenced by A's behavior as well as B's expectations, interpretations, and attributions with respect to A.

So there is no independent B sitting "out there" waiting to be impacted by A's message, as the old model would have it. Instead A and B are locked in *a relationship of simultaneous, mutual interdependence.*

Jervis illustrates the "interpretive traps" that can result from this double contingency, using the example of a leader issuing a statement about confidence in his or her abilities. Receivers of such a message know the leader is concerned about what inferences they might draw,

> [b]ut the very fact that everyone knows that these impressions are so important increases the chances that they will not be drawn in a straightforward way, thereby complicating matters for both actors and observers. Thus the latter may believe that acts which at first glance show confidence are likely to be taken only when the situation is desperate. . . . On the other hand, if audiences expect such a statement to be forthcoming, they may be even more alarmed by its absence. Furthermore, the notion of confidence itself is at least partly interactive in that the faith that one person can have in a leader is in part a function of his estimate of the confidence that others have.[25]

Another important aspect of complexity is that systems have emergent properties—the whole is more than the sum of its parts. It is impossible to reduce the success of a well-functioning work group, sports team or military unit to the skills or actions of any one member. Likewise, in our complex system the communication process is not completely under the control of either A or B. What they do matters, of course. But so does the action of the system *as a whole,* and it is in an important sense independent of the actions of the individual participants. The system is not necessarily under anyone's control.

One implication is that the system has effects of its own that can thwart the best intentions of its members. Even if a message is clearly sent and correctly decoded and received, it might still not create the desired interpretations and attributions. This is partly due to the effects of the double contingency. But as we explained in the previous section, when an interpretive system is in place it tends to assimilate new messages and reproduce itself. In this situation, the attempts to manage and control the message prescribed by the old message influence model are dysfunctional because repetition only serves to make the existing system stronger, and control is itself an action that is likely to be interpreted negatively.

A second implication is that the purpose of communication is not to cause acceptance and persuade the receiver to think in a particular way, as in the old model. In the PCOM framework the purpose of communication is to perturb the communication system and overcome its tendency to interpret and attribute in standard ways. This is especially true in conflict situations, where there are standard "recipes," "scripts," and "templates" for understanding the other party. As in the examples from the previous section, *any* conventional diplomatic message from the United States received by Muslims is likely to be interpreted as evidence that it does not understand them and is trying to impose its secular Western values. Only behavior that undermines the existing framework is likely to bring about a different response.

The purpose of communication is to perturb the system and overcome its tendency to interpret and attribute in standard ways.

Unfortunately it is not a simple task to envision a game-changing move because of the complexities of double contingency. This leads to our third implication, that *failure is the norm*. The message influence model assumes that unless there are debilitating levels of noise, or encoding/decoding processes distort the message, it will successfully travel from the source and implant itself in the mind of the receiver. But given PCOM assumptions, we can understand just how unlikely this scenario is. Interpretation by a receiver is influenced by an array of factors that are outside the control of—and may even be unknown to—the sender. Not the least of these is a system that is trying to preserve itself by resisting change.

Strategic communication is best viewed as an unpredictable and risky tool and should be used accordingly.

Table 2. Pragmatic Complexity Model Summary

Communication concept	Interpretation and attribution of the actions of system members
Constraint	Double contingency
Principles	• Control is impossible and dysfunctional • Less is more • Perturb stable system structures
Expectation	Failure

A final implication is that when it comes to strategic communication, *less is more*. The effects of messages are often unpredictable and may have delayed and indirect effects.[26] Thus there is risk in having too many messages in play before their impact is fully understood. Furthermore, messages potentiate both identification with, and division from, the intentions of the sender, leading to perverse effects: A message might increase understanding, but it might also create misunderstanding. Strategic communication is best viewed as an unpredictable and somewhat risky tool, and it should be used accordingly. The pragmatic complexity model is summarized in Table 2.

Living with Pragmatic Complexity

The model just discussed may seem discouraging, especially to those who have always believed in the validity of the message influence model. But to reiterate, we do not dismiss that model entirely. Communication still involves sending messages, and that is naturally done with the intent of influencing others. Doing this skillfully is still an admirable goal. But communicators should undertake their task with an updated 21st-century realism about what is actually happening in the process. That, we believe, is accurately described by the PCOM model. The model carries several principles for more effective communication, a few of which we outline now.

1. **Deemphasize control and embrace complexity.** A clear implication of the PCOM model of communication is: You can't control the message; get over it. The more we try to treat communication as a simple,

straightforward task with outcomes we can control, the less we are likely to succeed. Communication takes place in a complex system of double contingency that can be partially influenced but not controlled by the participants. Communicators should accept this reality and try to work with it, just as Wall Street traders accept the chaos of the market and try to "go with the flow." Once we let go of the idea of a well-ordered system that is under our control, we can start to think of what *is possible* in situations of uncertainty.

You can't control the message. Get over it.

For example, strategic communicators reading this paper have probably been thinking primarily about how the PCOM model constrains them and complicates their ability to operate effectively. Yet in the GWOT it is not just the West that is trying to have influence. The "bad guys" are trying to have their influence as well, and there is evidence that they too are applying an outdated strategy based on simple repeated messages.[27] This exposes them just as surely to the pitfalls of PCOM. If the West embraces complexity and they do not, this creates an asymmetry in its favor (for once). Among other things, it could use this asymmetry to seek ways to increase *its opponents'* exposure to negative consequences of unacknowledged complexity.

2. **Replace repetition with variation.** Unwavering use of a few simple messages is no more likely to work in a complex communication system than is a plan to always buy (and only buy) the same stock on Wall Street. What is needed in both cases is strategic experimentation. We advocate an evolutionary approach to sensemaking in the complex system, involving steps of variation, selection, and retention.[28] This approach is in the tradition of American pragmatists, like John Dewey,[29] who believed that democracy requires continual experimentation to discover the conditions under which social systems thrive.

Take an evolutionary approach involving variation, selection, and retention.

Rather than a grand overall strategy, communicators should rely on variations on a message theme. These are backed by small (rather than large) commitments and are followed up with careful observation of results. Communicators temporarily sustain things that "work" and perhaps add resources to them. They might also attempt further variations along the same lines. However they are agile, remaining ready to abandon existing messages at the first sign that they have lost their positive effects, switching to new variations.

For instance, rather than always promoting the virtues of democracy, the United States might try messages that discuss its problems and invite comparison of these faults to the problems of other forms of government. This could steer a conversation toward Churchill's famous conclusion that "democracy is the worst form of government, except all the others that have been tried." Doing this would reproduce Western values of freedom of thought and expression and show that we are not afraid of criticism. Another variation might be to argue that America "is the most Islamic country in the world" based on comparison between its values and those expressed in the Quran.[30] Again, unflattering comparisons to Muslim countries could be pointed out. These messages might work and they might not, but they are worth trying if backed by small, temporary commitments and careful observation of effects.

3. **Consider disruptive moves.** While variation can contribute to system change in an evolutionary sense, large-scale, transformative change typically only occurs only as a result of some major disruption in the normal operations of a system. There is no better recent example of this than the 9/11 attacks on the United States. Some critics argue that the United States over-reacted to the attacks.[31] But these arguments notwithstanding, it was clearly a game-changing event. International sympathy poured out for the United States.[32] More than 30% of the country changed its support in favor of President Bush.[33] The 9/11 Commission report speaks of a transformations in national priorities, government programs, and military strategies.[34] The attacks dramatically disrupted the structure of the existing system.

Transformative change only occurs as a result of a major disruption in the normal operations of a system.

While a disruption of this magnitude is a rare occurrence, its lesson is worth applying to strategic communication efforts. Since the structure of the communication system is not currently favorable to the West, we should consider disruptions that can change the game. The Mother of All Disruptions would be a breakthrough in the Israel-Palestine conflict. It goes without saying that U.S. support for Israel in this conflict complicates its relations with Arab countries. It also provides Al-Qaeda with ideological resources. In the most recent example, perceived U.S. foot-dragging regarding a cease-fire in the recent Israel-Hezbollah conflict provided the extremists with a rhetorical bonanza.[35] More generally,

> Al-Qaeda capitalizes on scenes of Israeli-Palestinian fighting which are widely disseminated through the Arab media on satellite channels like al-Manar and al-Jazeera. So even if Israeli policy is used instrumentally by al-Qaeda's leaders, the effectiveness of the tactics and the wider sympathy it generates depend on the state of relations between Israelis and Palestinians.[36]

A resolution to the Israel-Palestinian conflict would be a game-changer that would deny Al-Qaeda an important ideological tool and open up new possibilities for relations between Arab states and the West.

Though not as dramatic, another disruption that is sure to occur in 2009 is a change in the U.S. presidency. Some see rapid turnover in the Executive Branch caused by term limits as a liability. Be that as it may, it provides a regular possibility for disruption of international relationships and the rhetorical structure of strategic communication systems. The foreign policy of the current administration has attracted a good deal of international criticism, whether deserved or not, so the coming disruption has the potential for a significant impact. Planning should begin as soon as possible to capitalize on this opportunity.

4. **Expect and plan for failure.** The communication systems described by the PCOM model are complex. They contain multiple double contingencies, making it difficult to predict exactly what effects will result from particular messages. This means that, especially in terms of the "big picture," it is difficult to be strategic in the sense of setting a desired future state of affairs and mapping a set of logical steps that are likely to bring it about. Given our point above that well-intentioned efforts can have unanticipated perverse effects, it is perhaps just as likely that goals will be undermined as it is that they will be accomplished.

Think less in terms of grand strategy and more in terms of contingency planning.

With this in mind, strategic communicators should think less in terms of grand strategy and more in terms of contingency planning. Rather than assuming a message will be understood as it is intended, they should think of the ways things could go wrong, what the consequences of those outcomes will be, and the steps that might be undertaken in response. Then, if the message has the intended effects it is all to the good, and if it does not, options are immediately available for further variation as described above.

Conclusion

Current strategic communication practice in the United States and Western countries is based on an outdated message influence model from the 1950s that views communication as a process of transmission from a source to a receiver using simple, consistent, repeated messages. This model fails because it does not recognize communication as a meaning-making process. In reality, messages are interpreted within a large, complex system with emergent properties and self-preserving dynamics. The old model should be replaced with a 21st-century view of communication as interpretation and attribution of actions in an uncertain environment. Communicators are locked in simultaneous, mutual interdependence that reduces the value of grand strategy and makes failure the most likely outcome.

To succeed in this environment communicators should deemphasize control and embrace complexity, replace repetition of messages with experimental variation, consider moves that will disrupt the existing system, and make contingency plans for failure. Making these changes will create an asymmetry in the favor of the West, which it can exploit to great advantage in strategic communication aspects of the GWOT.

Notes

1. Corman, S.R. and Schiefelbein, J.S. (2006). *Communication and media strategy in the jihadi war of ideas*. Report #0601, Consortium for Strategic Communication, Arizona State University.
2. The design of messages is not the only aspect of this war of words. For example, the means used to communicate messages and the socio-cultural conditions that surround them are important too. But the messages themselves are nonetheless crucial, and the thinking behind message formulation matters a lot in the eventual success of communication efforts.
3. Berlo, D.K. (1960). *The process of communication: An introduction to theory and practice*. New York: Holt, Rinehart and Winston.

4. See Berlo, page vi. The ICAO was later absorbed into the U.S. Agency for International Development. According to the *Guide to Federal Records* it coordinated foreign aid but was also responsible for all non-military security programs.
5. Shannon, C. and Weaver, W. (1949). *The mathematical theory of communication*. Urbana, IL: University of Illinois Press.
6. See Berlo, op. cit. pages 30–32.
7. See Berlo, op. cit. Chapter 3.
8. Packard, V. (1957). *The hidden persuaders*. New York: David McKay Co.
9. Fawkes, J. (2001). Public relations and communications. In A. Theaker (Ed.), *The public relations handbook* (Chapter 2). London: Routledge.
10. The White House (2003, January 24). Establishing the Office of Global Communications. *Executive order 13282*, Office of Policy Coordination and International Relations.
11. Kean, T.H., et al. (2004). *The 9/11 Commission Report*, pp. 376–377.
12. Ford, J.T. (2004, August 23). *State Department and Broadcasting Board of Governors expand post-9/11 efforts but challenges remain*. Testimony before the Subcommittee on National Security, Emerging Threats, and International Relations, House Committee on Government Reform, p. 9.
13. Minow, N.M. (2003). *The whisper of America*. Foundation for Defense of Democracies, pp. 7–10.
14. Wolf, C.R. and Rosen, B. (2004). *Public diplomacy: How to think about it and improve it*. Occasional paper, Rand Corporation, p. 8.
15. Axley, S.R. (1984). Managerial and organizational communication in terms of the conduit metaphor. *Academy of Management Review, 9*, p. 431.
16. See Goodall, Jr., H.L. (2006). Why we must win the war on terror: Communication, narrative and the future of national security. *Qualitative Inquiry, 121, 30–59* and Goodall, Jr., H.L., Trethewey, A., & McDonald, K. (2006). *Strategic ambiguity, communication, and public diplomacy in an uncertain world*. Report #0604, Consortium for Strategic Communication, Arizona State University.
17. For a fuller account of Ms. Hughes's "listening tour," see *Diplomatic toast The New Republic* (2005, October 15).
18. Ibid.
19. *Remarks by the President* at the 20th Anniversary of the National Endowment for Democracy. United States Chamber of Commerce, November 6, 2003.
20. Combating Terrorism Center (2006). *Militant Ideology Atlas*. United States Military Academy at West Point.
21. Corman, S.R. (2006). Weapons of mass persuasion: Communicating against terrorist ideology. *Connections: The Quarterly Journal, 53,* 325–338.
22. Cited in Tannen, D. (1990). *You just don't understand: Women and men in conversation*. New York: Ballentine Books, p. 282.
23. Watzlawick, P. (1992). The construction of clinical 'realities.' In J. Zeig (Ed.). *The*

evolution of psychotherapry: The second conference. Brunner/Mazel, New York: Pearl, p. 64.

24. For the English translations see Luhmann, N. (1995). *Social systems*. Stanford University Press, Stanford, CA. (Originally published 1984). For an easier-to-read introduction, see Chaptcr 16 of Münch, R. (1994). *Sociological theory: From the 1850s to the present*. London: Burnham, Inc.
25. Jervis, R. (1997). *System effects: Complexity in political and social life*. Princeton, N.J.: Princeton University Press, p. 256. This book makes excellent further reading for the issues of complexity discussed in this paper.
26. Ibid. Chapter 2.
27. See Corman and Scheifelbein (2006), op. cit.
28. Weick, K.E. (1979). *The social psychology of organizing*, 2nd ed. Reading, MA: Addison-Wesley; Weick, K.E. (2006). Faith, evidencc, and action: Better guesses in an unknowable world. *Organization Studies, 27(11)*, 1723–1736.
29. Dewey, J. (1938). *Art as experience*. New York: Perigree.
30. For this argument we are indebted to Imam Mohamad Bashar Arafat of the Civilizations Exchange & Cooperation Foundation, personal communication, November 1, 2006.
31. For example, see Mueller, J. (2006). *Overblown: How Politicians and the terrorism industry inflate national security threats, and why we believe them*. New York: Free Press.
32. See the *September 11 News Page*.
33. Poll finds a united nation. *USA Today* (2001, September 16).
34. See the 9/11 Commission Report, Chapter 12.
35. See Corman (2006), op. cit.
36. Pressman, J. (2003). The primary role of the United States in Israeli-Palestinian relations. *International Studies Perspectives, 4 (2)*, 191–194.

CHAPTER NINE

Creating a New Communication Policy

How Changing Assumptions Leads to New Strategic Objectives

STEVEN R. CORMAN, ANGELA TRETHEWEY, AND H. L. GOODALL, JR.

Introduction

We conclude the book by looking at the U.S. National Strategy for Public Diplomacy and Strategic Communication (NSPDSC)[1] in light of the missed opportunities described in the first chapter and illustrated in the subsequent ones. We do not presume to have all the answers or pretend to know all the practical considerations in deploying such a strategy. However, we are convinced that there is value in viewing strategy through the lens of a more modern perspective on communication. The goal in doing this is to reveal an alternate point of view, grounded in radically different starting assumptions, which could create a badly needed disruption of the present system that frustrates Western objectives. We begin by identifying existing implicit assumptions of NSPDSC, and grounds for questioning them. Next we propose a set of alternative assumptions for thinking about U.S. communication efforts; then we conclude by proposing a new set of strategic objectives based on these.

Existing Assumptions

Surprisingly, the NSPDSC does not state its starting assumptions. The words "assume" and "assumption" never appear in the 34-page document. Yet reading between the lines we find important, if unstated, assumptions framing the plan's main recommendations. We begin by making six of these assumptions explicit.

1. **The United States is a credible source and can use its credibility to influence world public opinion.** This assumption underlies virtually the whole plan. It views the United States being a "champion" of values it holds dear, offering its "vision" to the world. It aims to "counter" ideologies that oppose this plan and to "promote" interests that conform to it. The U.S. cannot succeed at this unless it is an active, competent communicator that relevant audiences will trust—e.g., unless it registers the basic dimensions of credibility.[2] By assuming these qualities, it also assumes it has the persuasive power to do the championing, visioning, promoting, and countering envisioned in the plan. Accordingly, "United States Ambassadors," who are the most symbolic representatives of the government abroad, "should be the 'voice' of America" (p. 6).
2. **American values are cultural universals.** From the outset the plan is framed in the terms of classic American values like freedom of religion, equal justice, and human rights. But it goes beyond acknowledging these values as guiding principles by declaring them universals to which "all individuals, men and women, are equal and entitled" (p. 2). This assumption is also reflected in the explicit emphasis placed on education in the plan, the idea being that most of those who oppose us do so because they do not properly understand our values. The "diplomacy of deeds" section follows similar logic. It says that by actions we "can communicate our values and beliefs far more effectively than all of our words" (p. 7), and that the problem is primarily one of ignorance: "too few people (including those in our own country) know the tremendous impact Americans are making on lives around the world every day" (p. 7).
3. **The United States has worldwide partners who work at a common enterprise.** In a sense, this is a corollary of assumption 2. Since we all share universal values, we also have a common interest in working

together. Thus "the United States Government seeks to partner with nations and peoples across the world in ways that result in a better life for all of the world's citizens." (p. 2). The words "partner" and "partnership" occur 36 times in the NSPDSC, our partners being variously envisioned as other governments and nations, their peoples, centers of world power, the private sector, and local radio/television stations. Information is to be shared with our partners through the "State Department's 'Partnership for a Better Life' website" (p. 10).

4. **U.S. strategic communication efforts operate on discrete audience segments.** The plan has a section entitled "Strategic Audiences" that names five such audiences: "key influencers," youth, women/girls, minorities, and mass audiences. It envisions different messages being customized for each audience.
5. **Radio and television are the most important media.** Increasing numbers of people get their news and information from television and radio, according to the plan. In a section entitled "modernize communications" the plan concludes that "United States government officials must significantly expand their presence and appearances on foreign media" (p. 6). While the last part of this section says the U.S. also needs to improve its Internet presence, and later parts of the plan propose things like "digital outreach teams," it positions mainstream media as its main technological emphasis.
6. **Traditional, control-oriented methods of managing strategic communication will be effective if implemented properly.** In a sense, the previous assumptions do not represent any kind of change. The United States has always viewed itself as a credible partner in promoting universal values and has always targeted particular audiences using mass media. If our efforts are not working, it must be because we are not doing these things well enough. Thus the plan devotes an entire section to "Interagency Coordination" for the purpose of producing tighter central control over strategic communication activities, pursued through traditional methods.

Alternative Assumptions

Are these assumptions reasonable? Evidence suggests that the answer is no. Here are six alternative starting assumptions that in our view are better supported and provide the basis for a different approach to U.S. strategic communication.

1. **The United States has significantly diminished credibility and limited ability to influence world public opinion as a direct source.** In a paper published in early 2006,[3] we noted several studies showing a decline in perceptions of trustworthiness, competence, and goodwill, the main dimensions of credibility. Since then, evidence of this erosion has continued to mount. A recent poll by the Pew Global Attitudes Project[4] finds that anti-American attitudes remain consistent with levels of the previous five years, widespread distrust of American leadership, and "increasing disapproval of the cornerstones of U.S. foreign policy" (p. 5). Another poll[5] shows that the situation is particularly bad with respect to Arab countries. Without credibility, the U.S. can send all the messages it wants, but they will not have the intended effects. Zaharma sums up the situation this way: "What U.S. officials don't seem to register is that no amount of information pumped out by U.S. public diplomacy will be enough to improve the U.S. image. The problem, ultimately, is not lack of information but lack of credibility."[6]
2. **American values often conflict with those held by target audiences.** It is difficult for Americans to imagine others not sharing their values. We regard them as self-evident, timeless truths, and we believe that if everyone shared them the world would be a better place. One reservation is that even here in the United States, our belief in American values is not absolute. Since the 9/11 attacks, for example, we have lessened the value we place on freedom of speech and religion, separation of powers, and privacy. Today we allow the government to act in ways that would be been unthinkable before. Another point is that other cultures reject some of our values or interpret them in alternative ways, meaning they cannot serve as the basis for persuasive appeals. "Freedom" is a good example: An old saw from the Cold War is that Americans think about "freedom to" while the communist countries thought about "freedom from." Today we could substitute "Muslim" for "communist" in that sentence and it would be just as true. A third complication is that some of our values actively conflict with values from other cultures. An example of this, given earlier in the book, is our determination to promote democracy. Islamists view democracy as

a polytheistic (and therefore corrupt) religion, and our promotion of it as an attempt to destroy the Muslim community. The harder we try to promote such values, the more it works to our disadvantage. It is more realistic to assume that, despite their appeal to us, American values do not make a good basis for universal persuasive appeals.

3. **The United States will act unilaterally when necessary if doing so is in its interests.** While the United States may prefer to view itself as an equal partner in world affairs, it is safe to say that few of its interlocutors view us quite so benignly. Even Europe—the continent most like the U.S. in values and one with considerable and growing economic power—has had bitter clashes with the U.S. since 9/11. These include but are not limited to the Iraq War, rendition of terrorism suspects, and surveillance of airline passengers. In each of these cases the United States has ignored or over-ruled European reservations, launching the Iraq war, carrying out covert operations, and demanding airline passenger data. Relations with Russia, China, the Middle East, and other parts of the world are even less equal and cooperative. The point here is not to judge the propriety of these actions but simply to point out that they are not consistent with the actions of a "partner." Calling ourselves that in light of the reality that everyone else perceives compounds our credibility problems.
4. **Audiences for U.S. strategic communication are blurred.** It may be useful for analytic purposes to separate audiences for U.S. strategic communication, but as anything more than a thought exercise it is a dangerous move. Once upon a time it was safer to think of different audiences—domestic vs. international, those from different countries, different demographic groups—as discrete groups. Today, however, the lines are blurred by new media and globalization such that discrete messages designed for particular audiences easily escape to targets for which they were not intended. An example of this is President Bush's use of the word "crusade" to describe U.S. plans after the 9/11 attacks. His remarks, intended for a domestic audience, "passed almost unnoticed by Americans, [but] rang alarm bells in Europe"[7] and the Middle East, where (combined with other statements about battling evil) he was seen to be declaring a war of civilizations between Christians and Muslims. More recently, Western media reports have been used in enemy propaganda in Iraq.[8] It is easy to imagine how messages destined

for Muslim women (one of the NSPDSC targets) could be used by adversaries as evidence of U.S. intent to corrupt Muslim society.

5. **New media are the most important media.** There is little doubt that traditional media—based on a one-to-many mass communication model—are still important or that members of strategically significant audiences get information from them. But there is general consensus that they are being subsumed in the New Media whose chief feature is interactivity.[9] Traditional sources like radio and television are a part of this, but rather than passively receiving information through a limited number of channels, audience members today make active, individual choices about what messages to consume from a vast selection. Traditional broadcast sources make up only a small part of it. Web sites (many run by traditional media outlets), blogs, message boards, instant messages, and the like have an equal place in the mix. The system is interactive because audience members influence one another's choices: Think of how often in recent years you have sent or received a link about an interesting story via e-mail. At one time the problem was how to get a message into a closed media system. Today getting a message "out there" is easy; you only need to post it on a blog. The real challenge is making messages *sticky*. In the New Media environment, messages that are simple, unexpected, concrete, credible, emotional, and story-containing[10] are the ones people will remember and circulate through their personal networks. A splendid example is a short video called the Battle at Kruger that was posted on YouTube[11] (a heavy-hitting source in the New Media). It involves an epic battle between lions, crocodiles and water buffalo with an unexpected ending. It has been viewed (at the time of this writing) an incredible 16.7 million times as a result of the link being passed from person to person.
6. **Success in today's complex communication environment requires methods based on pragmatic complexity.** As we showed in Chapter 8, U.S. strategic communication is based on an outdated, 20th century *message influence* model of communication that has evolved into standard operating procedure in the last 60 years. It emphasizes being "on message," repetition, and control. This was effective in the "old media" environment, and it can still be effective in receptive, culturally homogeneous audiences. But in the current strategic communication environment its methods are counterproductive. There is a pressing need to

deemphasize control and embrace complexity, replace repetition with variation, consider moves that will disrupt the present system, and to do contingency planning for cases when messages do not have the desired effect.

New Strategic Objectives

Table 1 summarizes the NSPDSC's assumptions, and the contrasting assumptions just described. The alternate assumptions point to different ways of prosecuting the ongoing ideological struggle. Here we explain six new strategic objectives framed by our alternative assumptions.

In the Long Term the U. S. Must Work to Restore Lost Credibility

Lack of credibility is the key U.S. problem in our efforts to resist terrorist ideology. It limits our persuasive power and provides our enemies the means to easily discredit virtually any message we produce—even true and right ones. Our opponents need only to suggest that what we say cannot be trusted and/or that we do not comprehend the implications of what we are saying and/or that everything we say is based on some hidden motive to harm the target audience. Members of the audience are already predisposed to believe such claims, so they will be evaluated uncritically. In this situation the very possibility of effective strategic communication is implausible as best.

The first step toward solving any problem such as this is to acknowledge it. We cannot formulate a solution to our credibility deficit until we acknowledge it to ourselves, take it seriously, and orient our actions toward it. Admitting this problem to ourselves would put us in the mindset of working to resolve the problem instead of going on as if it does not exist. More important, we must acknowledge the problem publicly to our important strategic audiences. Behaving as though we have credibility and persuasive power when we do not is not only ineffective, it contributes to an image that we are "clueless" about the problem. This in itself contradicts one of the key dimensions of credibility, competence.

Table 1. Comparative Assumptions Underlying Public Diplomacy and Strategic Communication Policies.

NSPDSC Implicit Assumptions	Alternative Assumptions
The United States is a credible source and can use its credibility to influence world public opinion.	The United States has significantly diminished credibility and limited ability to influence world public opinion as a direct source.
American values are cultural universals.	American values conflict with values held by target audiences.
The United States has worldwide partners who work at a common enterprise.	The United States will act unilaterally when necessary if doing so is in its interests.
U.S. strategic communication efforts operate on discrete audience segments.	Audiences for U.S. strategic communication are blurred.
Radio and television are the most important media.	New Media are the most important media.
Traditional, control-oriented methods of managing strategic communication will be effective if implemented properly.	Success in today's complex communication environment requires methods based on pragmatic complexity.

A contrite admission that we have damaged our own credibility would not be easy or typical for the United States. Indeed it would take people by surprise, and so it would be an example of the kind of disruptive communication advocated in Chapter 8. Applying the principle of *counter-attitudinal advocacy*,[12] doubts about U.S. credibility would be acknowledged and used as a resource to develop common ground with the audience. This would signal that we are not "clueless," but rather that we "get it" and plan to take actions to redeem ourselves. It would have the effect of turning over a new leaf, putting our credibility-diminishing actions in the past, and beginning to undermine claims that everything we say and do is part of the same old pattern of behavior.

After admitting our credibility problems, we should focus on improving our image with respect to three specific dimensions of credibility, discussed here in order of importance. The first, *trustworthiness*, can be defined as *confidence in the reliability of a person or system*. The NSPDSC calls for a "diplomacy of deeds" (p. 7). However, it frames the problem as one of strategic audiences not understanding all the good things we do for the world. Doing good things is important for improving the third dimension of credibility (goodwill), but doing bad things at the same time simply makes us look fickle and unreliable. The same is true when we extol values like democracy but then react badly when democratic processes do not produce the results we want. To regain trustworthiness, there must be absolute consistency and alignment between what the U.S. does and what it says, even if audiences do not like what we are doing or saying. Above all it must be predictable in this way.

The second dimension of credibility, *competence*, can be defined as the *capability of a person or entity to carry out its plans*. Examples of how the U.S. has failed to carry out its plans in the global war on terrorism are legion. It has not captured Osama bin Laden "dead or alive"; it did not conclude the Iraq war in a matter of months or find weapons of mass destruction there, and it has not conducted itself in accordance with the values stated in the NSPDSC in the cases of Abu Ghraib and Guantanamo Bay. Even if we did not plan these outcomes and they do not result from malicious intent, in judgments of competence it is results that ultimately count. One way to reduce the impact of such cases is to be more circumspect in public statements about our goals: If the U.S. does not state forcefully what it plans to accomplish, then there can be no public failure to meet these plans. Conversely, an image of competence can be promoted by only announcing plans when there is virtual certainty that they will be accomplished.

The third dimension of credibility, *goodwill*, can be defined as *the perception that a person or entity takes actions that benefit others and avoid causing them harm*. As we already mentioned, the NSPDSC's "diplomacy of deeds" is a right step in helping re-establish this image. However, goodwill is an *on-balance* comparison that can easily be undone by actions demonstrating ill will on the other side of the ledger. To help counteract these events, the U.S. should dial up its efforts to do good. Goodwill is also the lest important of the three dimensions of credibility, meaning that even if we practically never harmed anyone, it would not be enough to overcome perceptions of mistrust and incompetence.

In the Shorter Term the U.S. Must Use Communication Techniques That Are Better Suited for a Low-Credibility Source

Restoring lost U.S. credibility is a long-term process, but strategic communication is still possible in the short term if it is based on methods that are suited for the low-credibility situation. One is to reduce the U.S *branding* on strategic messages. For example, rather than sourcing messages to high-level U.S. officials as is recommended in the NSPDSC, this function should be performed by lower-level officials and other Americans who have more day-to-day contact or something closer to peer relations with ordinary members of the target audience. A good step in this direction has already been taken by the State Department, with establishment of "digital outreach teams"[13] of lower level, Arabic-speaking employees who engage discussions on Arabic-language chat rooms and discussion boards.

Another option for low-credibility sources is to employ trusted third parties as messengers. One example of this is moderate American Muslims who oppose terrorist ideology. However they face risks in trying to deliver terrorism counter-narratives and not just from the terrorists. By participating on message boards and chat rooms frequented by the Bad Guys they face surveillance and possible targeting by the Department of Homeland Security, which monitors such traffic to identify potential terrorists. Addressing this concern might open up a new source of messages that would be in the U.S. strategic interest.

The U. S. Should Emphasize Common Interests in its Rhetoric Rather than Traditional American Values

If we dispense with the idea that American values are universal then it becomes clear that they are not suitable as the bedrock of U.S. strategic communication efforts. For one thing, as we have explained already, our values are actually in conflict with those of other cultures. This undermines their effectiveness as argument sources, and emphasizing them can be used against us. For another, our values become tainted by the same general credibility problems described above. If American values underlie American behavior, and this behavior is viewed as untrustworthy, incompetent, and ill willed, then the value of the values is called into sharp question.

A better basis for our strategic communication would be shared interests. Values are abstract and derive from deep-seated cultural norms and beliefs.

Interests, on the other hand, are more concrete and practical and are linked to shorter-term goals of specific groups. People can have the same interests in particular situations even if they do not share the same values overall, as is routinely demonstrated in political action. Framing our commonalities with others in terms of interests rather than values moves from an cultural/ideological frame to a political/practical one that is less likely to encounter resistance.

The U. S. Must Replace Partnership Language with Collaboration Language and Declare Interests it is Willing to Defend Unilaterally

The word partnership implies a long-term, stable relationship that is based on closely aligned interests and a willingness to compromise and always create a united front. The United States does have long-term stable relationships with some countries, especially those in Europe. But recent history has shown a de-alignment of interests even with them. As we move into other areas of the world and down in scale, our relations become even less partner-like. In the Middle East, with the possible exception of Israel, relationships are better described as short term and changing, with only partial alignments of interests. In Iraq we have recently developed relationships with smaller-scale groups with whom only months ago we were fighting heated battles. It is also apparent to everyone that in certain cases we will not hesitate to "go it alone" no matter what our allies think.

Making heavy use of partnership language in cases like this makes the U.S. seem disingenuous and compounds our credibility problems. Specifically it creates expectations that our behavior does not meet and, combined with our perceived military and economic power, this diminishes our trustworthiness. A solution to this problem is to rhetorically lower expectations for our co-action with other countries and groups that is better aligned with what we actually do.

We propose *collaboration* as a replacement for the current U.S. partnership language. This word carries the same sense of working jointly on a problem but does not carry the relational implications that go with partnership. It also aligns with the previous objective of shifting to a shared-interests framework, because collaborations can and do occur between groups that are loosely connected but come together on a temporary basis to exploit some opportunity. Collaboration is more consistent with what we actually do, and using this rhetorical frame is likely to improve perceptions of trustworthiness.

Likewise we believe it is vital that we better articulate situations in which we are willing to act unilaterally. As we argued in Chapter 2, strategic ambiguity is a useful thing, but it is possible to have too much of a good thing. Currently the U.S. position is that it will act alone when doing so is "in its national interests." But there is no consensus even in the U.S., let alone in other countries, about what its national interests are. Such a broad and moving target allows almost any situation to be defined as relating to a vital national interest. This makes us seem unpredictable and therefore untrustworthy to our potential collaborators. While we do not advocate anything like a set of "rules," the U.S. should formulate more explicit policy about exactly what its national interests are and make it clear to other countries that these will always constitute grounds for breaking collaborations.

The U.S. Should Develop a Working Model of Complex Audience/Media Systems

The old days when it was possible to target messages at a foreign audience without worry that they would impact domestic audiences are long gone. Today domestic messages "leak" to foreign countries and between target audiences in foreign countries. Likewise mass media have been subsumed in the New Media. Mainstream media messages now circulate on the Internet, and the mass media report on things that happen online.

These blurry boundaries between audiences and the ongoing integration of the media mean the U.S. must stop thinking of any of these entities as distinct, independent, and manageable. Instead audiences and media should be viewed as part of a *complex adaptive system*. As defined by Holland,[14]

> A Complex Adaptive System (CAS) is a dynamic network of many agents (which may represent cells, species, individuals, firms, nations) acting in parallel, constantly acting and reacting to what the other agents are doing. The control of a CAS tends to be highly dispersed and decentralized. If there is to be any coherent behavior in the system, it has to arise from competition and cooperation among the agents themselves. The overall behavior of the system is the result of a huge number of decisions made every moment by many individual agents.

Thinking of a complex adaptive system in place of discrete audiences and channels will force the U.S. to focus on the interdependence of these entities and how they will interact. With a working model of these interdependencies, it will be better able to exercise caution when formulating messages and avoid negative unintended consequences and mixed signals that result when messages reach audiences for which they were not intended.

The U.S. Must Develop a New Model of Strategic Communication Management Consistent with Pragmatic Complexity Principles

Not only must the U.S. think of audiences and media as a complex system, but as we proposed in Chapter 8, the entire strategic communication enterprise must be reconceptualized in these terms. Once messages reach audiences through whatever channel, they become part of a complex system of social construction, governed by the constraints of the double contingency. This makes old-style attempts to "control the message" futile.

In place of control-oriented strategies the U.S. should embrace complexity and attempt to anticipate and guide the flow of discourse, analogous to the way successful stock traders deal with a complex market. It should replace the old-style practice of repeating a small number of predetermined messages with an approach that deploys varied messages in an attempt to find the ones that "stick." Change can only occur if there is movement, and patterns of response in the existing system tend to resist this. So the U.S. must also consider disruptive moves designed to jolt the system out of these institutionalized patterns of response. Because of unpredictability of the system, it must also lower expectations for the ability of any one approach or strategy to change the game. An evolutionary approach of patient variation, selection, and retention is the one most likely to yield success.

Finally, organizational changes are needed to support a pragmatic complexity approach. Analysts have repeatedly noted how the hierarchical organization of the government, its "stovepipes," and the separation of strategic communication from policy hinder its ability to be effective in resisting terrorist ideology. To effectively deploy weapons of mass persuasion, this must change. As Gregory has aptly noted: "Strategic direction for public diplomacy must be flexible and adaptive. Its models should be connectors and networks, not stovepipes and hierarchies."[15]

Conclusion

We began the book by describing four missed opportunities in the struggle against terrorist ideology. The first of these was reliance on a vocabulary that is inappropriate for the situation. The language of war is inappropriate because you cannot go to war with a noun and hope to have a decisive victory. As Chapter 7 pointed out, there is no territory to be taken, no definitive end, and no treaty-signing ceremony on a ship. Yet the war framing calls for the narratives that create just these kinds of expectations.

This is but one of the many important language issues. As we explained in this chapter, use of the term "partnership" for U.S. relations with other countries is not accurate and creates unrealistic expectations that in the end harm its credibility. Deploying language that is heavily influenced by American values causes important members of the target audience to resist these messages. As we have argued throughout the book, language is crucial in constructing the reality of what the U.S. is doing and in helping guide others' perceptions. Words matter, and to act otherwise misses an opportunity for positive persuasion.

A second missed opportunity is to engage in a productive dialog with allies and adversaries. U.S. strategic communication since 9/11 has been seen, almost universally, as one-sided and heavy-handed—an effort to manipulate world opinion. To be sure, that is an accurate reading in that it represents the approach the U.S. has taken, using advertising and marketing techniques to sell its "brand." Meanwhile it has forgone opportunities for dialog (for example, in the Ahmadinejad letter described in Chapter 5) treating communication as a scarce resource and making itself appear afraid to talk. As we argued in this chapter, there is a need to transcend this kind of thinking with a model based on engagement and dialog.

A third missed opportunity is that we are letting the Bad Guys beat us at our own game. As we explained in Chapter 4, this is in part because they have a focused, coherent communication strategy that they execute in ways that take advantage of important asymmetries. But for the most part the U.S. has created its own shortcomings. As described in this chapter, it has allowed its credibility to erode while acting as if this were not the case. Partly because of this it has been forced into actions, like providing martyrdom images of al Zarqawi, that have handed the Bad Guys communicative advantages. As we recommended in chapter 2, the U.S. needs to seek "unified diversity" based on cooperation in place of "focused wrongness" based on dominance and power.

The fourth missed opportunity, in a way, encapsulates the others. It is the failure to make use of the power of narrative. The Bad Guys deploy a powerful and interesting narrative of an oppressive force resisted by them, the heroic, holy warriors. As we noted in Chapter 4, this narrative is key to their religious legitimation and the only thing that keeps them from being seen as murderers and thugs. To the extent that there is a coherent U.S. narrative, it is that we are trying to bring our values, like democracy, to the Muslim world. Unfortunately this is the very nature of the oppression being claimed in the counter-narrative. The U.S. should seek to change the subject by de-emphasizing its values and shifting the story toward the death and suffering in the Muslim community caused by the Bad Guys and their ideology.

Toward these ends, we have shown in this chapter that it is possible to reject the old assumptions and start with new ones. When we do this a whole new set of strategic objectives emerges that begin to address the missed opportunities. The previous chapters have outlined several communication principles *in situ.* Taken together, these principles (e.g., strategic ambiguity, pragmatic complexity, leveraging networks, narratives, and vocabulary) can guide the construction of new ways of engaging the global struggle against violent extremism and formulating viable objectives and policy for strategic communication.

Moreover, the new strategic objectives must shape and inform a new and innovative strategic communication policy. That policy would be far less concerned with controlling and reinforcing a core message about U. S. values than with using the creativity and openness of those values to perturb violent extremists and their audiences. The point of the perturbance would be to get individuals enmeshed within loosely connected but powerful social networks to reconsider the nature and value of their own ideological commitments. This new strategic communication policy would be less occupied with staying on message and repeating it, than with creating opportunities for diverse and even conflicting homegrown narratives that create doubt about the credibility of violent extremists' promises and unwavering trust in what they represent as "truth." In so doing, the new policy would create networked space to empower local meanings capable of enabling homegrown change. Finally, it would be a policy that takes full advantage of contemporary communication understandings that recognize that meanings are in social networks, not in words.

Notes

1. U.S. Department of State (2007, May). U.S. national strategy for public diplomacy and strategic communication. Retrieved August 14 from http://uscpublicdiplomacy.org/pdfs/stratcommo_plan_070531.pdf.
2. Corman, S.R., Hess, A., & Justus, Z.S. (2006). Credibility in the Global War on Terrorism: Strategic Principles and Research Agenda. Report #0603, Consortium for Strategic Communication Arizona State University. Available at http://co-

mops.org/article/117.pdf.

3. Ibid.
4. Pew Global Attitudes Project (2007, June). Global Unease With Major World Powers. Accessed September 8, 2007 at http://pewglobal.org/reports/pdf/256.pdf
5. Zogby International (2006, December). Five nation survey of the Middle East. Accessed September 8, 2007 at http://aai.3cdn.net/96d8eeaec55ef4c217_m9m6b-97wo.pdf
6. Zaharma, R.S. (2006, December). The U.S. Credibility Deficit. Foreign Policy in Focus. Accessed September 8, 2007 at http://www.fpif.org/fpiftxt/3796
7. Europe cringes at Bush "Crusade" Against Terrorists (2001, September 19). *Christian Science Monitor*. Accessed September 11, 2007 at http://www.csmonitor.com/2001/0919/p12s2-woeu.html
8. Propaganda Wars 2 (2007, September 7). Protein Wisdom Blog. Accessed September 11, 2007 at http://proteinwisdom.com/?p=9754
9. See Crosbie, V. (2006, April) What is 'New Media'? Corante Blog. Accessed September 13, 2007 at http://rebuildingmedia.corante.com/archives/2006/04/27/what_is_new_media.php
10. Heath, C., & Heath, D. (2007). *Made to stick: Why some ideas survive and others die*. New York: Random House.
11. *http://www.youtube.com/watch?v=LU8DDYz68kM*. Thanks to John Rendon for this example.
12. Stiff, J. B., & Mongeau, P. A. (2003). *Persuasive communication*, 2nd ed. New York: Guilford Press.
13. Described in the NSPDSC. See also McFarquhar, N. (2007, September 22). At State Dept., Blog Team Joins Muslim Debate. *New York Times*.
14. Waldrop, M. M. (1993). *Complexity: The emerging science at the edge of order and chaos*. New York: Lawrence Sternlieb, p. 145.
15. Gregory, B. (2005). Public diplomacy and strategic communication: Cultures, firewalls, and imported norms. Paper presented at the American Political Science Association Conference on International Communication and Conflict, Washington, DC. Available online at http://www8.georgetown.edu/cct/apsa/papers/gregory.pdf

Useful Links

9/11 Commission
http://www.9 11commission.gov

ASU Center for the Study of Religion and Conflict
http://www.asu.edu/csrc

ASU Consortium for Strategic Communication
http://comops.org

Brookings Institution

Diplomacy Page
http://www.brookings.edu/topics/diplomacy.aspx

Terrorism Page
http://www.brookings.edu/topics/terrorism.aspx

Central Intelligence Agency "Strategic Intent 2007–2011"
https://www.cia.gov/about-cia/strategic-intent-2007–2011.html

COMOPS Journal
http://comops.org/journal

Dipnote Blog
http://blogs.state.gov/

GWU Public Diplomacy Institute
http://www.pdi.gwu.edu

Harmony and Disharmony: Exploiting al-Qa'ida's Organizational Vulnerabilities (pdf):
http://smallwarsjournal.com/documents/ctc1.pdf

Heritage Foundation Public Diplomacy Page
http://www.heritage.org/research/publicdiplomacy

Jamestown Foundation
http://jamestown.org

Kim Andrew Elliott
http://kimelli.nfshost.com/

MountainRunner Blog
http://mountainrunner.us/

Phil Taylor's Web Site
http://ics.leeds.ac.uk/papers/index.cfm?outfit=pmt

Public Diplomacy Blog
http://uscpublicdiplomacy.com/index.php/newsroom/pdblog_main

Rand Corporation Terrorism and Homeland Security Page
http://www.rand.org/pubs/online/terrorism/

Rand Report on Marketing and Strategic Communication
http://www.rand.org/pubs/monographs/2007/RAND_MG607.pdf

Small Wars Journal
http://smallwarsjournal.com/

St. Andrews University Centre for the Study of Terrorism and Political Violence
http://www.st-andrews.ac.uk/~wwwir/research/cstpv/

Terrorism Analysts
http://www.terrorismanalysts.com

"The Making of a Martyr" Documentary by Brooke Goldstein (download for free)
http://movies.aol.com/truestories/making-of-a-martyr

University of Southern California Center for Public Diplomacy
http://uscpublicdiplomacy.org

U.S. Air University Strategic Communication Page
http://www.au.af.mil/info-ops/strategic.htm

U.S. Department of Homeland Security
http://www.dhs.gov

U.S. Department of State
http://www.state.gov

Under Secretary for Public Diplomacy and Public Affairs
http://www.state.gov/r/

Office of the Coordinator for Counterterrorism
http://www.state.gov/s/ct/

U.S. Federal Bureau of Investigation Counterterrorism Page
http://www.fbi.gov/terrorinfo/counterrorism/waronterrorhome.htm

U.S. National Strategy for Public Diplomacy and Strategic Communication (June, 2007).
http://www.state.gov/documents/organization/87427.pdf

U.S. News and World Report Article on the Pentagon's Strategic Communication Plan:
http://www.usnews.com/usnews/news/articles/060529/29propaganda.htm

West Point Combating Terrorism Center
http://ctc.usma.edu

Appendix

Text of Iranian President's Letter to President Bush

EUP20060509394001 Paris* Le Monde *in English 09 May 06

[Text of letter written by Iranian President Ahmadinejad to U.S. President Bush; *Le Monde* provided no information on how the letter was obtained]

Mr George Bush, President of the United States of America

For sometime now I have been thinking, how one can justify the undeniable contradictions that exist in the international arena—which are being constantly debated, specially in political forums and amongst university students. Many questions remain unanswered. These have prompted me to discuss some of the contradictions and questions, in the hopes that it might bring about an opportunity to redress them.

Can one be a follower of Jesus Christ (PBUH), the great Messenger of God? Feel obliged to respect human rights, Present liberalism as a civilization model, Announce one's opposition to the proliferation of nuclear weapons and WMDs, Make War and Terror his slogan, And finally, Work towards the establishment of a unified international community—a community which Christ and the virtuous of the Earth will one day govern, But at the same time, Have countries attacked; The lives, reputations and possessions of people destroyed and on the slight chance of the . . . of a . . . criminals in a village city, or convoy for example the entire village, city or convey set ablaze. Or because of the possibility of the existence of WMDs in one country, it is occupied, around one hundred thousand people killed, its water sources, agriculture and industry destroyed, close to 180,000 foreign troops put on the ground, sanctity of private homes of citizens broken, and the country pushed back perhaps fifty years. At what price? Hundreds of billions of dollars spent from the treasury of one country and certain other countries and tens of thousands of young men and women—as occupation troops—put in harm's way, taken away from family and loved ones, their hands stained with the blood of others, subjected to so much psychological pressure that every day some commit suicide and those returning home suffer depression, become sickly and grapple with all sorts of aliments; while some are killed and their bodies handed to their families.

On the pretext of the existence of WMDs, this great tragedy came to engulf both the peoples of the occupied and the occupying country. Later it was revealed that no WMDs existed to begin with.

Of course Saddam was a murderous dictator. But the war was not waged to topple him, the announced goal of the war was to find and destroy weapons of mass destruction. He was toppled along the way towards another goal, nevertheless the people of the region are happy about it. I point out that throughout the many years of the . . . war on Iran Saddam was supported by the West.

Mr. President,

You might know that I am a teacher. My students ask me how can these actions be reconciled with the values outlined at the beginning of this letter and duty to the tradition of Jesus Christ (PBUH), the Messenger of peace and forgiveness.

There are prisoners in Guantanamo Bay that have not been tried, have no legal representation, their families cannot see them and are obviously kept in a strange land outside their own country. There is no international monitoring of their conditions and fate. No one knows whether they are prisoners, POWs, accused or criminals.

European investigators have confirmed the existence of secret prisons in Europe too. I could not correlate the abduction of a person, and him or her being kept in secret prisons, with the provisions of any judicial system. For that matter, I fail to understand how such actions correspond to the values outlined in the beginning of this letter, i.e. the teachings of Jesus Christ (PBUH), human rights and liberal values.

Young people, university students and ordinary people have many questions about the phenomenon of Israel. I am sure you are familiar with some of them.

Throughout history many countries have been occupied, but I think the establishment of a new country with a new people, is a new phenomenon that is exclusive to our times.

Students are saying that sixty years ago such a country did not exist. The show old documents and globes and say try as we have, we have not been able to find a country named Israel.

I tell them to study the history of WWI and II. One of my students told me that during WWII, which more than tens of millions of people perished in, news about the war, was quickly disseminated by the warring parties. Each touted their victories and the most recent battlefront defeat of the other party. After the war, they claimed that six million Jews had been killed. Six million people that were surely related to at least two million families.

Again let us assume that these events are true. Does that logically translate into the establishment of the state of Israel in the Middle East or support for such a state? How can this phenomenon be rationalized or explained?

Mr President,

I am sure you know how—and at what cost—Israel was established: Many thousands were killed in the process.

Millions of indigenous people were made refugees.

Hundred of thousands of hectares of farmland, olive plantations, towns and villages were destroyed.

This tragedy is not exclusive to the time of establishment; unfortunately it has been ongoing for sixty years now.

A regime has been established which does not show mercy even to kids, destroys houses while the occupants are still in them, announces beforehand its list and plans to assassinate Palestinian figures and keeps thousands of Palestinians in prison. Such a phenomenon is unique—or at the very least extremely rare—in recent memory.

Another big question asked by people is why is this regime being supported? Is support for this regime in line with the teachings of Jesus Christ (PBUH) or Moses (PBUH) or liberal values? Or are we to understand that allowing the original inhabitants of these lands—inside and outside Palestine—whether they are Christian, Muslim or Jew, to determine their fate, runs contrary to principles of democracy, human rights and the teachings of prophets? If not, why is there so much opposition to a referendum?

The newly elected Palestinian administration recently took office. All independent observes have confirmed that this government represents the electorate. Unbelievingly, they have put the elected government under pressure and have advised it to recognize the Israeli regime, abandon the struggle and follow the programs of the previous government.

If the current Palestinian government had run on the above platform, would the Palestinian people have voted for it? Again, can such position taken in opposition to the Palestinian government be reconciled with the values outlined earlier? The people are also saying why are all UNSC resolutions in condemnation of Israel vetoed?

Mr President,

As you are well aware, I live amongst the people and am in constant contact with them—many people from around the Middle East manage to contact me as well. They dot not have faith in these dubious policies either. There is evidence that the people of the region are becoming increasingly angry with such policies.

It is not my intention to pose too many questions, but I need to refer to other points as well.

Why is it that any technological and scientific achievement reached in the Middle East regions is translated into and portrayed as a threat to the Zionist regime? Is not scientific R&D one of the basic rights of nations?

You are familiar with history. Aside from the Middle Ages, in what other point in history has scientific and technical progress been a crime? Can the possibility of scientific achievements being utilized for military purposes be reason enough to oppose science and technology altogether? If such a supposition is true, then all scientific disciplines, including physics, chemistry, mathematics, medicine, engineering, etc. must be opposed.

Lies were told in the Iraqi matter. What was the result? I have no doubt that telling lies is reprehensible in any culture, and you do not like to be lied to.

Mr President,

Don't Latin Americans have the right to ask, why their elected governments are being opposed and coup leaders supported? Or, why must they constantly be threatened and live in fear?

The people of Africa are hardworking, creative and talented. They can play an important and valuable role in providing for the needs of humanity and contribute to its material and spiritual progress. Poverty and hardship in large parts of Africa are preventing this from happening. Don't they have the right to ask why their enormous wealth—including minerals—is being looted, despite the fact that they need it more than others?

Again, do such actions correspond to the teachings of Christ and the tenets of human rights?

The brave and faithful people of Iran too have many questions and grievances, including: the coup d'etat of 1953 and the subsequent toppling of the legal government of the day, opposition to the Islamic revolution, transformation of an Embassy into a headquarters supporting, the activities of those opposing the Islamic Republic (many thousands of pages of documents corroborates this claim), support for Saddam in the war waged against Iran, the shooting down of the Iranian passenger plane, freezing the assets of the Iranian nation, increasing threats, anger and displeasure vis-à-vis the scientific and nuclear progress of the Iranian nation (just when all Iranians are jubilant and collaborating their country's progress), and many other grievances that I will not refer to in this letter.

Mr. President,

September Eleven was a horrendous incident. The killing of innocents is deplorable and appalling in any part of the world. Our government immediately declared its disgust with the perpetrators and offered its condolences to the bereaved and expressed its sympathies.

All governments have a duty to protect the lives, property and good standing of their citizens. Reportedly your government employs extensive security, protection and intelligence systems—and even hunts its opponents abroad. September Eleven was not a simple operation. Could it be planned and executed without coordination with intelligence and security services—or their extensive infiltration? Of course this is just an educated guess. Why have the various aspects of the attacks been kept secret? Why are we not told who botched their responsibilities? And, why aren't those responsible and the guilty parties identified and put on trial?

All governments have a duty to provide security and peace of mind for their citizens. For some years now, the people of your country and neighbors of world trouble spots do not have peace of mind. After 9.11, instead of healing and tending to the emotional wounds of the survivors and the American people—who had been immensely traumatized by the attacks—some Western media only intensified the climates of fear and insecurity—some constantly talked about the possibility of new terror attacks and kept the people in fear. Is that service to the American people? Is it possible to calculate the damages incurred from fear and panic?

American citizens lived in constant fear of fresh attacks that could come at any moment and in any place. They felt insecure in the streets, in their place of work and at home. Who would be happy with this situation? Why was the media, instead of conveying a feeling of security and providing peace of mind, giving rise to a feeling of insecurity?

Some believe that the hype paved the way—and was the justification—for an attack on Afghanistan. Again I need to refer to the role of media. In media charters, correct dissemination of information and honest reporting of a story are established tenets. I express my deep regret about the disregard shown by certain Western media for these principles. The main pretext for an attack on Iraq was the existence of WMDs. This was repeated incessantly—for the public to, finally, believe—and the ground set for an attack on Iraq.

Will the truth not be lost in a contrived and deceptive climate? Again, if the truth is allowed to be lost, how can that be reconciled with the earlier mentioned values? Is the truth known to the Almighty lost as well?

Mr. President,

In countries around the world, citizens provide for the expenses of governments so that their governments in turn are able to serve them.

The question here is what has the hundreds of billions of dollars, spent every year to pay for the Iraqi campaign, produced for the citizens?

As your Excellency is aware, in some states of your country, people are living in poverty. Many thousands are homeless and unemployment is a huge problem. Of course these problems exist—to a larger or lesser extent—in other countries as well. With these conditions in mind, can the gargantuan expenses of the campaign—paid from the public treasury—be explained and be consistent with the aforementioned principles?

What has been said, are some of the grievances of the people around the world, in our region and in your country. But my main contention—which I am hoping you will agree to some of it—is: Those in power have specific time in office, and do not rule indefinitely, but their names will be recorded in history and will be constantly judged in the immediate and distant futures.

The people will scrutinize our presidencies.

Did we manage to bring peace, security and prosperity for the people or insecurity and unemployment? Did we intend to establish justice, or just supported especial interest groups, and by forcing many people to live in poverty and hardship, made a few people rich and powerful—thus trading the approval of the people and the Almighty with theirs'? Did we defend the rights of the underprivileged or ignore them? Did we defend the rights of all people around the world or imposed wars on them, interfered illegally in their affairs, established hellish prisons and incarcerated some of them? Did we bring the world peace and security or raised the specter of intimidation and threats? Did we tell the truth to our nation and others around the world or presented an inverted version of it? Were we on the side of people or the occupiers and oppressors? Did our administration set out to promote rational behavior, logic, ethics, peace, fulfilling obligations, justice, service to the people, prosperity, progress and respect for human dignity or the force of guns. Intimidation, insecurity, disregard for the people, delaying the progress and excellence of other nations, and trample on people's rights? And finally, they will judge us on whether we remained true to our oath of office—to serve the people, which is our main task, and the traditions of the prophets—or not?

Mr. President,

How much longer can the world tolerate this situation? Where will this trend lead the world to? How long must the people of the world pay for the incorrect decisions of some rulers? How much longer will the specter of insecurity—raised from the stockpiles of weapons of mass destruction—hunt the people of the world? How much longer will the blood of the innocent men, women and children be spilled on the streets, and people's houses destroyed over their heads? Are you pleased with the current condition of the world? Do you think present policies can continue?

If billions of dollars spent on security, military campaigns and troop movement were instead spent on investment and assistance for poor countries, promotion of health, combating different diseases, education and improvement of mental and physical fitness, assistance to the victims of natural disasters, creation of employment opportunities and production, development projects and poverty alleviation, establishment of peace, mediation between disputing states and distinguishing the flames of racial, ethnic and other conflicts where would the world be today? Would not your government and people be justifiably proud? Would not your administration's political and economic standing have been stronger? And I am most sorry to say, would there have been an ever increasing global hatred of the American governments?

Mr. President, it is not my intention to distress anyone.

If Prophet Abraham, Isaac, Jacob, Ishmael, Joseph or Jesus Christ (PBUH) were with us today, how would they have judged such behavior? Will we be given a role to play in the promised world, where justice will become universal and Jesus Christ (PBUH) will be present? Will they even accept us?

My basic question is this: Is there no better way to interact with the rest of the world? Today there are hundreds of millions of Christians, hundreds of millions of Moslems and millions of people who follow the teachings of Moses (PBUH). All divine religions share and respect one word and that is monotheism or belief in a single God and no other in the world.

The holy Koran stresses this common word and calls on all followers of divine religions and says: [3.64] Say: O followers of the Book! Come to an equitable proposition between us and you that we shall not serve any but Allah and (that) we shall not associate aught. With Him and (that) some of us shall not take others for lords besides Allah, but if they turn back, then say: Bear witness that we are Muslims. (The Family of Imran).

Mr. President,

According to divine verses, we have all been called upon to worship one God and follow the teachings of divine prophets. To worship a God which is above all powers in the world and can do all He pleases. The Lord which knows that which is hidden and visible, the past and the future, knows what goes on in the Hearts of His servants and records their deeds. The Lord who is the possessor of the heavens and the earth and all universe is His court planning for the universe is done by His hands, and gives His servants the glad tidings of mercy and forgiveness of sins. He is the companion of the oppressed and the enemy of oppressors. He is the Compassionate, the Merciful. He is the recourse of the faithful and guides them towards the light from darkness. He is witness to the actions of His servants, He calls on servants to be faithful and do good deeds, and asks them to stay on the path of righteousness and remain steadfast. Calls on servants to heed His prophets and He is a witness to their deeds. A bad ending belongs only to those who have chosen the life of this world and disobey Him and oppress His servants. And a good and eternal paradise belongs to those servants who fear His majesty and do not follow their lascivious selves.

We believe a return to the teachings of the divine prophets is the only road leading to salvations. I have been told that Your Excellency follows the teachings of Jesus (PBUH), and believes in the divine promise of the rule of the righteous on Earth.

We also believe that Jesus Christ (PBUH) was one of the great prophets of the Almighty. He has been repeatedly praised in the Koran. Jesus (PBUH) has been quoted in Koran as well; [19:36] And surely Allah is my Lord and your Lord, therefore serves Him; this is the right path, Marium (Mary mother of Jesus).

Service to and obedience of the Almighty is the credo of all divine messengers.

The God of all people in Europe, Asia, Africa, America, the Pacific and the rest of the world is one. He is the Almighty who wants to guide and give dignity to all His servants. He has given greatness to Humans.

We again read in the Holy Book: The Almighty God sent His prophets with miracles and clear signs to guide the people and show them divine signs and purity them from sins and pollutions. And He sent the Book and the balance so that the people display justice and avoid the rebellious.

All of the above verses can be seen, one way or the other, in the Good Book as well.

Divine prophets have promised: The day will come when all humans will congregate before the court of the Almighty, so that their deeds are examined. The good will be directed towards Haven and evildoers will meet divine retribution. I trust both of us believe in such a day, but it will not be easy to calculate the actions of rulers, because we must be answerable to our nations and all others whose lives have been directly or indirectly affected by our actions.

All prophets, speak of peace and tranquility for man—based on monotheism, justice and respect for human dignity.

Do you not think that if all of us come to believe in and abide by these principles, that is, monotheism, worship of God, justice, respect for the dignity of man, belief in the Last Day, we can overcome the present problems of the world—that are the result of disobedience to the Almighty and the teachings of prophets—and improve our performance?

Do you not think that belief in these principles promotes and guarantees peace, friendship and justice?

Do you not think that the aforementioned written or unwritten principles are universally respected?

Will you not accept this invitation? That is, a genuine return to the teachings of prophets, to monotheism and justice, to preserve human dignity and obedience to the Almighty and His prophets?

Mr. President, History tells us that repressive and cruel governments do not survive. God has entrusted the fate of man to them. The Almighty has not left the universe and humanity to their own devices. Many things have happened contrary to the wishes and plans of governments. These tell us that there is a higher power at work and all events are determined by Him.

Can one deny the signs of change in the world today? Is this situation of the world today comparable to that of ten years ago? Changes happen fast and come at a furious pace.

The people of the world are not happy with the status quo and pay little heed to the promises and comments made by a number of influential world leaders. Many people around the world feel insecure and oppose the spreading of insecurity and war and do not approve of and accept dubious policies.

The people are protesting the increasing gap between the haves and the have-nots and the rich and poor countries.

The people are disgusted with increasing corruption.

The people of many countries are angry about the attacks on their cultural foundations and the disintegration of families. They are equally dismayed with the fading of care and compassion. The people of the world have no faith in international organizations, because their rights are not advocated by these organizations.

Liberalism and Western style democracy have not been able to help realize the ideals of humanity. Today these two concepts have failed. Those with insight can already hear the sounds of the shattering and fall of the ideology and thoughts of the liberal democratic systems.

We increasingly see that people around the world are flocking towards a main focal point—that is the Almighty God. Undoubtedly through faith in God and the teachings of the prophets, the people will conquer their problems. My question for you is: Do you not want to join them?

Mr President,

Whether we like it or not, the world is gravitating towards faith in the Almighty and justice and the will of God will prevail over all things.

Vasalam Ala Man Ataba'al hoda
Mahmood Ahmadi-Najad President of the Islamic Republic of Iran

Index

A

B

C

D

I

N

O

P

Q

R

S

T

U

V

W

ABOUT THE EDITORS

Steven R. Corman is Professor at Arizona State University where he directs the Consortium for Strategic Communication in the Hugh Downs School of Human Communication. A recognized expert on complex organizations, social networks, and new media, since 2001 he has served as an invited participant and featured speaker at numerous national and international symposia on counterterrorism and public diplomacy. He serves as a consultant to the Department of Defense and the Department State on issues of strategic communication and was a member of the Scientist Panel for the Strategic Operations Working Group at U.S. Special Operations Command.

Angela Trethewey is Associate Professor and Associate Director in the Hugh Downs School of Human Communication at Arizona State University, and is a member of the Consortium for Strategic Communication. Her scholarship centers on issues of ideology and power as they impact strategic communication processes. She is currently working on a Department of Defense sponsored project on self-organizing systems and armies of the future. She is co-author of the best-selling book *Organizational Communication: Balancing Creativity and Constraint, 5th ed.*

H. L. (Bud) Goodall, Jr. is Professor and Director of the Hugh Downs School of Human Communication at Arizona State University. He is the author and co-author of twenty books and over one hundred articles, chapters, and presentations. He is the recipient of the "Best Book of 2007" award from the National Communication Association. He serves as a U.S. Department of State International Speaker on countering ideological support for terrorism and improving public diplomacy and serves as a consultant on strategic communication to the Department of State and Department of Defense groups.